NEUROANATOMY

NEUROANATOMY

K. E. MOYER
Carnegie-Mellon University

HARPER & ROW, PUBLISHERS, New York
Cambridge, Hagerstown, Philadelphia, San Francisco,
London, Mexico City, São Paulo, Sydney

1817

Sponsoring Editor: George A. Middendorf
Project Editor: Pamela Landau
Designer: T. R. Funderburk
Production Manager: Kewal K. Sharma
Compositor: American–Stratford Press, Inc.
Printer and Binder: Halliday Lithograph Corporation
Art Studio: Vantage Art, Inc.

NEUROANATOMY

Library of Congress Cataloging in Publication Data

Moyer, Kenneth Evan, Date–
    Neuroanatomy.

    Includes indexes.
    1. Neuroanatomy.   I. Title.
QM451.M69   611'8   79-20773
ISBN 0-06-044639-0

To
Dusty
with love

CONTENTS

PREFACE

This book was designed to be useful for students of medicine, nursing, dentistry, psychobiology, and all of the sciences that require a knowledge of neuroanatomy. It has been organized in such a way that the student can get a good overview of the central nervous system and then work with that knowledge until it is well committed to memory and can serve as a useful internal reference whenever neuroanatomical concepts are encountered.

It has also been organized for easy review so that students may return to the text and study particular sections or concepts needed for a given task. The figures and the labels are uniquely arranged to provide ready access to the information. Thus, if students come across an unfamiliar term in their other reading, they can quickly locate it on a diagram in this book so that it can be understood in relation to the rest of the nervous system. Therefore, this book is valuable not only as a text but also as a convenient reference tool.

The pedogogical concepts on which this book was based were tested repeatedly in different classes. I am grateful to the many students who used early versions of this material with drawings that were all too crude. These students made many helpful comments that were incorporated into the final text.

The usual acknowledgement reserved for secretaries is inadequate for Wanda Keppler. Wanda did much more than type this book, keep the files, arrange the drawings, and struggle with my sometimes undecipherable script. She has put her stamp on this work in many ways. She read the material carefully and called to my attention inconsistencies, obscure passages, and confusing statements. Her work has been tireless, faithful, and ever cheerful. It has been a real pleasure to work with her.

Thanks are also due to Dr. Michael Crabtree who was kind enough to go through the entire manuscript. He caught many errors and made some valuable suggestions.

Finally, I must acknowledge Dusty's help with this book as with the ones that have come before it. She has helped the most by being here and being loving.

K. E. Moyer

NEUROANATOMY

INTRODUCTION

A Note to the Student

This book was designed not just to provide you with information on neuroanatomy, but to facilitate your learning it as easily as possible. The format provides a new approach to learning. There is no easy road to gaining knowledge of the nervous system, but it is clear that some methods are easier than others. As most students know by this time in their careers, the passive reading of material does not promote attention and acquisition. Most of you, at one time or another, have read an entire chapter in a textbook only to realize an hour later that you could not recall a single particle of the contents, whereas you could recall in detail, the drifting, shifting kaleidoscope of memories and daydreams that occupied your mind while you were actually reading the text word for word. It is clear that some kind of active study is needed in which the mind is too engaged in the study process to escape into the more pleasant and attractive recesses of the mind where the hopes, dreams, and fond memories are kept. One approach to the facilitation of active study was the programmed text, in which small portions of the material were presented in the so-called frames and then followed by a blank in which material learned from a previous frame was to be written. The programmed text provided for active study, for it is difficult to daydream while filling blanks. However, for any particular student, the programmed text was almost always wrong. The material was either presented so fast that the student got lost, or it was presented so slowly that it was necessary repeatedly to go over material already well committed to memory. This book has been designed to involve you in active study at a pace that is particularly suited to you.

One of the most difficult aspects of learning neuroanatomy is to achieve a good three-dimensional cognitive map of many complex structures in the central nervous system, to get a feeling for the shapes of the individual structures and their relationships with one another. Aside from dissection one of the best ways to acquire this cognitive map is to study in detail a large number of figures or diagrams from many different angles. When you can identify the same structure in a number of different drawings depicting different aspects of that structure and its relationship to other structures, you will gradually be able to fit this complexity into a unified whole. Then, when you encounter a neuroanatomical name in your reading, you will immediately be able to "see" that structure and its relationship to other structures in the brain.

How to Use This Book

With few exceptions, the text and figures are on alternate pages of this book. Whenever possible, the textual material on the left-hand page refers to the diagrammatical material on the right-hand page. Thus, as you read a particular portion of the text you can readily refer to the related figure to reinforce your understanding with a visual representation of the same information. In general, you need not search for a remote figure in order to find an illustration of the concept under discussion. As you read the text you should refer frequently to the figures to get a better understanding of the relationship of one structure to another. The techniques used here make it easy for you to find a specific anatomical unit. All first citations of anatomical names are given in the text in **bold** type and are followed by an item or label number and a figure number, for example, **central fissure** (3–4). In each figure all of the labels are on the right-hand side and are numbered consecutively from the top to the bottom of the page. Thus, if you are reading about the **central fissure** (3–4), you can readily locate it by finding label 3 in Figure 4, and following the identification line to the sulcus itself.

Unfortunately, reading about a structure and finding it on a figure does not insure that you will remember it or that you will be able to identify it at some later time. In order to remember it, you must obviously locate it frequently on the figure and *recall* its name or label. The number of trials required to commit a given name to memory will vary tremendously from one individual to another depending on aptitudes and previous experience. It is also true that some individuals for idiosyncratic reasons will be able to remember one structure or set of structures more readily than some other individuals will. This book is designed to provide you with the amount of practice that *you* need, no more and no less.

You can study the text material and find on the figure the items described in the text. Then, in order to fix these materials in your mind, you should study the figure until you are able to identify each of the structures. When you think that you have the material well enough in mind, you can readily test yourself and get practice in recalling the names at the same time. Place a blank sheet of paper over the labels, leaving the label numbers exposed. Then write in the correct names on the blank sheet. When you finish, you can slide the page with your answers over and check to see how well you have done. Some students have found it easier to fold under the label portion of the figure page. The practice sheet is then placed under the figure portion of the page and the anatomical names are written in at the appropriate numbers.

You can repeat this process on each figure as many times as necessary to achieve the level of mastery you have set for yourself. For a variety of reasons, some names are much easier to learn than others. With this system you can concentrate on those labels with which you are having difficulty. After you have had a number of alternating studying and self-testing times and still find it difficult to recall the names of certain items, concentrate your studying on those alone. Do not hesitate to mark up your book and check the difficult items for further, more intensive study. The next few times that you self-test, do not take the time to write in the labels of all the terms you already know. Write out only the ones that are difficult for you.

There is considerable experimental evidence that long-term retention of this type of material is attained most readily by a combination of massed and distributed practice. Although the curve of forgetting is quite steep on original learning, it becomes less so on each subsequent relearning. It is therefore reasonable to practice enough to fill in the labels on a particular diagram to a level of 100 percent accuracy. Once you have achieved this level of mastery, go on to other figures or to other endeavors. Several hours or days later (according to what learning procedures are best for you) return to that figure and test yourself again until you achieve the same level of accuracy. Again, concentrate on the names with which you are having difficulty. Continue this process until you can, after significant intervals, achieve 100-percent mastery on the first trial.

It is a good idea to keep a record of your trials so that you can keep informed of your progress. The record need not be elaborate; the date and the number of errors or omissions on a given trial can be jotted down on a blank portion of the label page. You will find that this task becomes easier as you get farther along in the book. First, learning to learn is an important variable, and as you go along you will pick up techniques that will facilitate your learning. Second, many of the structures run through large segments of the nervous system and will be seen again and again. They will become old friends,

readily identified. As your cognitive map develops, they will be immediately recognized as being in the places where they were meant to be.

If you follow this procedure for all of the figures in this book, you will finish the course with an excellent three-dimensional cognitive map of the nervous system, which will serve you well through a wide variety of courses. Further, if at some later date you need to do so, you can return to this book and quickly bring yourself back to a very high level of mastery.

This text will also serve as a convenient reference because of the procedure used in indexing and because of the extensive use of figures. There are two indexes, one for the text and one for the figures. When, in your further studies, you encounter a neuroanatomical term about which you need more information, you will find it in one of the indexes. The index to the text will provide you with the pages on which the item is discussed. The index to figures will list those figures on which the structure can be found a well as the label number for the structure. Thus, the **red nucleus** (18–29) will be readily found as label 18 in Figure 29. It is not necessary to search through thirty-five or forty labels on a figure to find the item you want.

One of the problems that students encounter in the study of neuroanatomy is that of multiple names for the same structure. In this book, the preferred name will be indicated in **boldface** type. Additional names for a particular structure will be given and will be printed in *italics*. Generally the more descriptive name is preferred. For example, **middle cerebellar peduncle** is preferred over *brachium pontis.*

It is important for you to have some knowledge of the alternate terms. Two study aids have been incorporated in this book to help you gain familiarity with them. In the Appendix of Alternate Terms there is a list of alternate terms. Each term is listed in alphabetical order with the other terms. Thus, you will find

middle cerebellar peduncle = *brachium pontis*

and

brachium pontis = **middle cerebellar peduncle.**

The preferred term is in **boldface** whether it is listed first or last. By referring to the Appendix of Alternate Terms you will be able to find the various terms that are applied to a given structure. You will also be able to tell which of them is used most frequently in this book.

On the figures, an * is placed in front of the label of any structure that has more than one name for it. This is to remind you of that fact. If you do not remember the alternate names, you can look them up in the text, or in the Appendix of Alternate Terms.

Some terms have been used interchangeably throughout the book. Thus, tract and fasciculus are equivalent terms as are midbrain and mesencephalon.

Organization of this Book

The first section of the book is devoted to the gross external anatomical structure of the central nervous system. After a discussion of terms of orientation and an overview of the various divisions of the nervous system we start with the most obvious external feature of the nervous system, the **cerebral hemispheres** and work our way down to the tip of the **spinal cord**, the **filum terminale**. The next major section deals with the internal structure of the nervous system. In this section we start with the **spinal cord** and work our way up to the **cerebral hemispheres**. Then, in shorter sections we deal with the **menges** and **cerebrospinal fluid**, the **cranial nerves**, and the **autonomic nervous system**. By the time you cover that material, you will have a good idea of how the whole mechanism is put together and will be prepared to deal with the functional systems. First the afferent systems are covered and then the efferent systems. Finally, a section is devoted to a number of complex functional systems, including such mechanisms as sleep, temperature regulation, and food consumption.

A Note to the Instructor

This book, used in different ways, can meet the needs of several types of courses. It can be used as the basic text in a brief course in neuroanatomy when supplemented with the usual lectures and other materials. For longer, more-detailed courses, it can be used as a valuable supplement to longer and more-complete textbooks. The student is certain to benefit from the active study approach around which this book is designed.

There are a number of courses such as physiological psychology and neurobiology in which a knowledge of neuroanatomy is essential for an understanding of the material though it is not the major thrust of the subject matter. Further, many of these courses are offered at a level at which it is not feasible to require a course in

neuroanatomy as a prerequisite. This problem is frequently resolved by devoting several precious weeks of the semester to neuroanatomy or by allowing the students to learn about it on a haphazard basis. With its self-paced, self-taught format this book can help to alleviate the problem. There is more neuroanatomy in this text than most students need to know in order to understand the material in such courses as psychobiology. However, the format permits the instructor readily to screen out the less important details so that students can concentrate on the essentials.

Each anatomical structure on each figure is numbered. This feature makes it possible for the instructor to go through the text and cross out the items considered less important to the demands of a particular course or to a specific approach to a course. A list of figure and structure numbers considered as nonessential can then be given to the students who can cross off the items in their books with a yellow (transparent) felt-tip pen. They can then be confident that they will not be tested on those items. At the same time their books remain a useful reference if they encounter a particular structure in the literature and wish to look it up.

With this text students can be given the task of learning the three-dimensional "geography" of the nervous system on their own with little class time. The assignment can be defined by the pages to be covered and the item numbers considered important, and students can be told to return when they can pass a test on the material. In many courses, the figures in the text can serve as the testing device. The students can be given copies of the figures considered most relevant and asked to identify a particular sample of the numbered structures. The testing can be as exhaustive as desired.

The mastery concept is gaining considerable acceptance in educational circles today and has much to recommend it. It is the concern of professors that students have a certifiable knowledge of the material. They are less concerned with how long within the confines of the course structure it takes them to get to that point. Students are permitted to pace themselves and to take an examination of particular amounts of material whenever they are ready and as frequently as necessary to demonstrate a level of mastery determined by the professor. This textbook, used as a testing device, is particularly well suited to the self-paced mastery approach to teaching neuroanatomy.

The book can be divided into reasonable sections and

the student informed that he must demonstrate a given level of knowledge on each section. As described earlier, the tests consist of copies of the figures with a sample of the numbers on the figures that are to be identified. It is a simple matter to prepare a large number of different forms of the same test on any given section by merely selecting a different sample of the numbered labels that have been assigned. Since the student never knows which version of the test he will get, it is necessary to study all of the material and not work toward passing a particular examination. This process also means that a very large number of essentially equivalent forms of the same test are readily available. The student can then be tested on the same material as many times as necessary to demonstrate mastery. One need only insure that a given test form will not be given to the student more than once.

The tests can be easily administered and scored by a technician with a relatively low level of sophistication. It is also possible to provide the student with immediate feedback on how well he did on the examination regardless of how long it takes the technician to get around to grading the test. If carbon paper is used when taking the test, the student can take the carbon copy with him and quickly compare his answers with those in the book to determine whether he has mastered the material or must return to his studies.

SECTION 1

ORIENTATION

Top and *bottom, back* and *belly, side* and *center* are all common terms to indicate the relative positions of objects, items, or parts of the body. In neuroanatomy special terms are used to locate structures. The nomenclature for humans and animals, however, is slightly different because humans have taken to their hind legs. The belly area of the poodle in Figure 1 is referred to as the **ventral** (26–1) area, which is oriented toward the ground, while the spinal area is referred to as the **dorsal** (22–1) area, which is oriented upward. However, if this were a performing poodle walking around on its hind legs, the ventral and dorsal areas would be in the same relative positions as the ventral (26–1) and dorsal (22–1) areas of the young lady in Figure 1. A further source of confusion exists in the terms *anterior* and *posterior.* In animals such as our poodle **anterior** (21–1) refers to the front, or head, end and **posterior** (19–1) to the tail, or back, end. In humans, however, the front of a standing person is the same as the belly. Thus, in humans **anterior** (18–1) refers to the same relative direction as ventral, and **posterior** (19–1) refers to the same direction as dorsal.

Another minor source of confusion lies in the fact that for most animals the neural axis, from the tip of the spinal cord to the frontal poles of the brain, is essentially a straight line. Thus, the underside of the brain has the same orientation as the underside of the belly and is properly referred to as ventral. In humans, however, the enormous development of the **cerebral hemispheres** (12–1) has caused the nervous system to fold over at the top in a 90-degree bend. As a result, the top of the brain is really dorsal, and the base of the brain is ventral. These potential sources of confusion will become clear as you study the diagrams on the facing page.

Additional terms are used to clarify the relative positions of the parts of the nervous system. The direction toward the front of the brain is referred to as **cranial** (15–1), **superior** (1,14,20–1), or **rostral** (15,21–1), while the opposite direction, toward the tip of the spinal cord, is called **caudal** (7,17,23–1) or **inferior** (2,16,24–1). The term **medial** (6,9–1) indicates near or toward the midplane, while **lateral** (3,8–1) indicates the direction away from the midplane, or out toward the side. **Proximal** (9–1) means near or toward the central nervous system, while **distal** (8–1) means toward the periphery away from the central nervous system.

Another considerable help in understanding the relationship of the various parts of the nervous system stems from the division of the brain into three imaginary planes. If a slice is cut through the brain that divides it into right and left parts, the plane exposed will be the **sagittal** (11–1) plane. If that slice is precisely in the middle so that the two halves of the brain are of equal size, the exposed area will be the **median,** or *midsagittal,* plane. The **coronal,** or *frontal,* (10–1) plane divides the brain into front and back parts. It runs, of course, at right angles to the sagittal. Another term for the coronal plane is **transverse** (25–1) plane. This term is used particularly in reference to sections of the spinal cord exposed by slicing through the cord as if one were slicing bologna. A third plane runs at right angles to both the sagittal and coronal planes and is called the **horizontal** (13–1) plane.

Figure 1

Directional terms used in describing the nervous system.

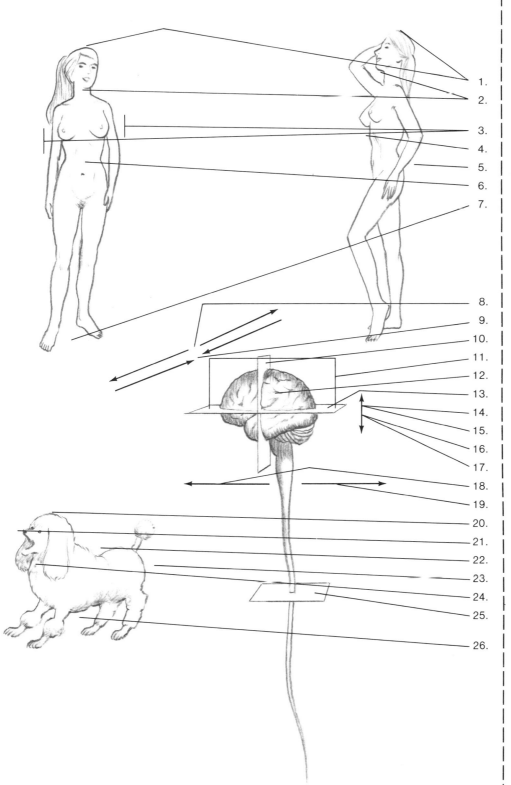

1. Superior or rostral
2. Ventral or inferior, refers to head only
3. Lateral
4. Anterior or ventral
5. Posterior or dorsal
6. Medial
7. Caudal

8. Lateral or distal
9. Medial or proximal
10. Coronal or transverse plane
11. Sagittal plane
12. Cerebral hemispheres
13. Horizontal plane
14. Dorsal or superior
15. Cranial or rostral
16. Ventral or inferior
17. Caudal
18. Anterior
19. Posterior
20. Superior
21. Anterior or rostral
22. Dorsal
23. Caudal
24. Inferior
25. Transverse plane of spinal cord
26. Ventral

SUBDIVISIONS OF THE NERVOUS SYSTEM

Nature did not divide the central nervous system into neat, mutually exclusive subdivisions. It is a continuous, functioning entity from the tip of the attachment of the **filum terminale** (4,11–14) to the **frontal pole** (1–3) of the brain. However, those who study neuroanatomy have found it convenient to refer to larger or smaller segments of the nervous system in order to specify anatomical locations and relationships. The arbitrary divisions discussed in this section and illustrated in Figure 2 are derived, in part, from the developmental sequence during the embryonic stage.

The brain develops from the dorsal hollow neural tube, which is derived from the middorsal ectoderm. Quite early in its development the brain shows three clearly identifiable enlargements as a result of rapid growth of cells around the neural tube. These are the most basic divisions of the central nervous system and consist of the **forebrain** (1–2) or *prosencephalon,* the **midbrain** (4–2), or **mesencephalon,** and the **hindbrain** (6–2), or *rhombencephalon.* Further development results in additional subdivisions, which are conventionally identified as belonging together. Thus, the three major subdivisions of the brain increase to five major subdivisions. The forebrain is divided into the **telencephalon** (2–2) and the **diencephalon** (3–2); the **midbrain** remains undivided as the **mesencephalon** (4–2); and the hindbrain is divided into the **metencephalon** (5–2) and the **myelencephalon** (7–2). As the brain continues to develop, additional structures can be identified. (See Table

1.) The neural tube also develops and expands differentially along its longitudinal axis and forms the spinal canal and the four ventricles of the brain. Each of these structures is associated with the main divisions of the brain (also indicated in Table 1).

Some of these terms are used more frequently than others, but you will encounter all of them in your further reading about the nervous system. It is well to learn these subdivisions thoroughly because they help provide an overall picture of the relationships of the major portions of the brain. As you learn more details, they will fit neatly into a coherent whole. The same system should be used in the study of the tables as is used in the study of the diagrams. The table is set up so that the right-hand columns can be covered with a piece of paper and the various divisions written in next to the numbers. The paper can then be moved to the right to check the answers and the spelling.

TABLE 1

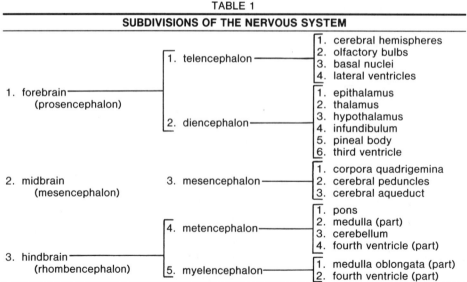

SUBDIVISIONS OF THE NERVOUS SYSTEM

1. forebrain (prosencephalon)	1. telencephalon	1. cerebral hemispheres 2. olfactory bulbs 3. basal nuclei 4. lateral ventricles
	2. diencephalon	1. epithalamus 2. thalamus 3. hypothalamus 4. infundibulum 5. pineal body 6. third ventricle
2. midbrain (mesencephalon)	3. mesencephalon	1. corpora quadrigemina 2. cerebral peduncles 3. cerebral aqueduct
3. hindbrain (rhombencephalon)	4. metencephalon	1. pons 2. medulla (part) 3. cerebellum 4. fourth ventricle (part)
	5. myelencephalon	1. medulla oblongata (part) 2. fourth ventricle (part)

Figure 2

Subdivisions of the brain.

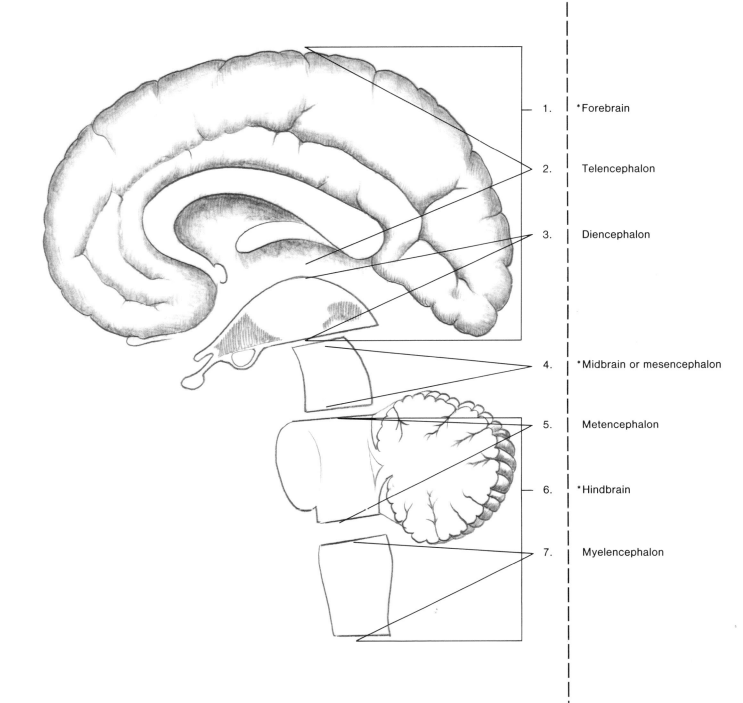

1. * Forebrain

2. Telencephalon

3. Diencephalon

4. * Midbrain or mesencephalon

5. Metencephalon

6. * Hindbrain

7. Myelencephalon

THE CEREBRAL HEMISPHERES: LATERAL ASPECT

The **cerebral hemispheres** are the most obvious structures of the nervous system. Their complexity provides the differences between animals and humans. In the next few sections we will look at the hemispheres from the lateral, medial, and ventral views.

In the lateral aspect (Fig. 3) the two hemispheres are incompletely separated along the midline by the **longitudinal fissure**. This deep cleft is bounded at the bottom by the **corpus callosum** (2,15,16–5), the large band of fibers that serves as the principal connection between the two hemispheres. The anterior part of the hemispheres is the **frontal pole** (1–3), and the posterior end as the **occipital pole** (23–3).

The **cerebral cortex** is spread over the area of the hemispheres, and its surface is greatly increased by being folded into convolutions called **gyri** (sing., **gyrus**), which are separated by grooves, or furrows, called **sulci** (sing., **sulcus**) or **fissures**. The general pattern of the gyri and sulci is essentially the same for all individuals, although there may be slight variations.

The **lateral fissure** (22–3), or *fissure of Sylvius,* is a deep, major sulcus that courses along the lateral surface of the hemisphere and divides the **frontal lobe** (1–4) and **parietal lobe** (2–4) above the fissure from the **temporal lobe** (10–4) below it. The most anterior portion of the temporal lobe is known as the **temporal pole** (37–3). The lateral fissure divides into three branches. The first branch, the **anterior horizontal ramus** (24–3), runs rostrally into the frontal lobe, and the second, the **anterior ascending ramus** (21–3), enters that lobe dorsally. The longest portion, the **posterior ramus** (19–3), as the name implies, courses backward toward the occipital pole and turns dorsally into the parietal lobe.

The next most obvious landmark on the cerebral hemispheres is the **central fissure** (10–3), or *fissure of Rolando.* It begins at the upper margin of the longitudinal cerebral fissure and runs obliquely across the dorsolateral surface of the hemisphere at about a 70-degree angle. The fissure ends just above and slightly behind the anterior ascending ramus of the lateral fissure. The area rostral to the central fissure and above the lateral cerebral fissure is the frontal lobe.

Although there is some variation, three principal sulci can be identified in the frontal lobe. The **precentral sulcus** (5–3) runs more or less parallel to the central fissure (10–3) and is sometimes divided into superior and inferior sections. The **superior frontal sulcus** (3–3) and the **inferior frontal sulcus** (12–3) originate in the precentral sulcus and extend rostrally along the curvature of

the dorsal portion of the hemisphere.

The sulci of the frontal lobe form and define the gyri. The **precentral gyrus** (11–3) lies just anterior to the central fissure and is bordered anteriorly by the precentral sulcus. The **superior frontal gyrus** (4–3), the **middle frontal gyrus** (7–3), and **inferior frontal gyrus** (20–3) are formed by the superior frontal sulcus and the inferior frontal sulcus. The two anterior rami of the **lateral fissure** (22–3) further subdivide the inferior frontal gyrus into three parts. The **orbital** (2–3) part is located just rostral to the anterior horizontal ramus; the **triangular** (38–3) part is wedge shaped and lies between the two anterior rami of the lateral fissure. Finally, the **operculum** (17–3) is situated between the precentral sulcus and the anterior ascending ramus of the lateral fissure.

The **temporal lobe** (10–4), which is ventral to the lateral fissure, is composed of three gyri defined by the sulci of that lobe. The **superior temporal gyrus** (26–3) is bounded dorsally by the lateral fissure and ventrally by the **superior temporal sulcus** (36–3). That sulcus then forms the dorsal boundary of the **middle temporal gyrus** (35–3), with the inferior boundary of that gyrus formed by the **middle temporal sulcus** (30–3). Finally, the **inferior temporal gyrus** (31–3) is bounded by the middle temporal sulcus and the inferior temporal sulcus.

The **parietal lobe** (2–4) extends from the central fissure posteriorly to the **parieto-occipital sulcus** (18–3). The **postcentral sulcus** (8–3) divides the parietal lobe into the **postcentral gyrus** (6–3) and the **parietal lobule.** The **intraparietal sulcus** (16–3) runs at right angles to the postcentral sulcus and separates the **superior parietal lobule** (9–3) above it from the **inferior parietal lobule** (15–3) below it. The inferior parietal lobule is further divided into the **supramarginal gyrus** (13–3), which curves around the posterior end of the lateral fissure, and the **angular gyrus** (14–3), which curves around the **superior temporal sulcus** (36–3).

The **occipital lobe** (8–4) occupies only a small portion of the dorsolateral surface of the hemisphere. The anterior boundary of this triangular area is formed by a line that joins the **parieto-occipital sulcus** (18–3) and the **preoccipital notch** (28–3). The most obvious fissure in the occipital lobe is the **calcarine fissure** (25–3). Most of this fissure lies on the medial surface (17,18–5) of the hemisphere, but the posterior end of it curves around onto the lateral surface, dividing the occipital lobe into superior and inferior aspects. In some individuals there is an additional sulcus that curves around the occipital pole called the **lunate sulcus** (27–3).

Figure 3

Lateral aspect of the cerebral hemispheres.

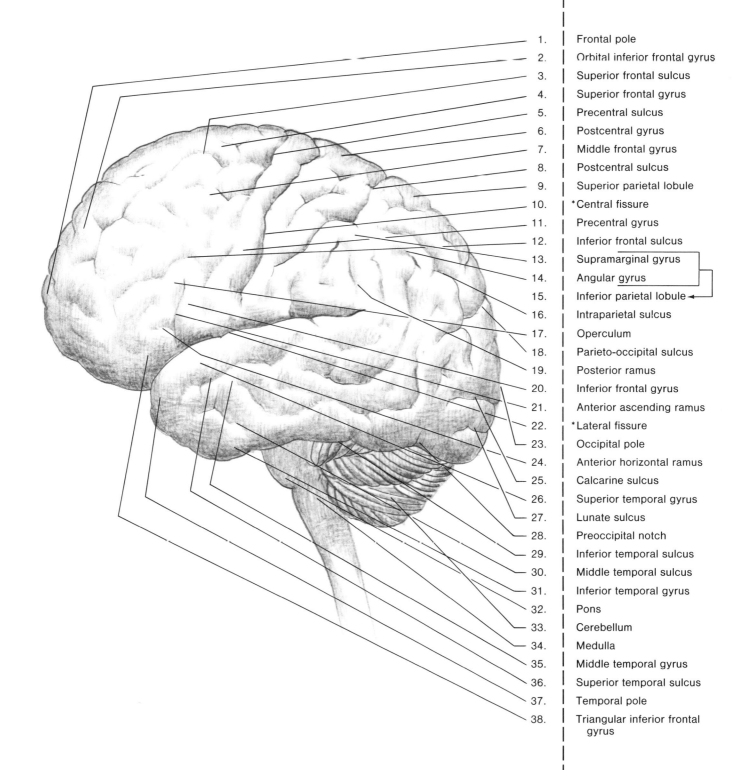

1. Frontal pole
2. Orbital inferior frontal gyrus
3. Superior frontal sulcus
4. Superior frontal gyrus
5. Precentral sulcus
6. Postcentral gyrus
7. Middle frontal gyrus
8. Postcentral sulcus
9. Superior parietal lobule
10. *Central fissure
11. Precentral gyrus
12. Inferior frontal sulcus
13. Supramarginal gyrus
14. Angular gyrus
15. Inferior parietal lobule
16. Intraparietal sulcus
17. Operculum
18. Parieto-occipital sulcus
19. Posterior ramus
20. Inferior frontal gyrus
21. Anterior ascending ramus
22. *Lateral fissure
23. Occipital pole
24. Anterior horizontal ramus
25. Calcarine sulcus
26. Superior temporal gyrus
27. Lunate sulcus
28. Preoccipital notch
29. Inferior temporal sulcus
30. Middle temporal sulcus
31. Inferior temporal gyrus
32. Pons
33. Cerebellum
34. Medulla
35. Middle temporal gyrus
36. Superior temporal sulcus
37. Temporal pole
38. Triangular inferior frontal gyrus

THE LOBES OF THE BRAIN AND THE INSULA

The lobes of the brain were discussed in Section 3. Figure 4a strips away the clutter of detail and provides the landmarks for ready identification. The **frontal lobe** (1–4), which includes about one-third the surface of the hemisphere, extends from the **frontal pole** (4–4) to the **central fissure** (11–4), or *fissure of Rolando*. The inferior border of the frontal lobe consists of the trunk and the posterior branch of the **lateral fissure** (6–4), or *fissure of Sylvius*.

The **parietal lobe** (2–4) extends from the central fissure back to the **parieto-occipital fissure** (5–4). As the figure indicates, the posterior boundary of the parietal lobe is not completely defined by the parieto-occipital fissure. Convention has determined that the complete boundary consists of an arbitrary line from that fissure to a small indentation on the ventrolateral surface of the temporal lobe about 4 centimeters from the occipital pole known as the **preoccipital notch** (9–4). The boundary between the temporal lobe and the parietal lobe also consists of an imaginary line continuing the lateral fissure to the line that forms the parietal lobe's posterior boundary.

The lateral fissure forms the superior border of the **temporal lobe** (10–4), and the posterior border is the imaginary line from the parieto-occipital fissure to the preoccipital notch. That same line forms the anterior border of the **occipital lobe** (8–4), which occupies the most posterior portion of the hemisphere. The rounded apex of that lobe constitutes the **occipital pole** (7–4).

THE INSULA, OR ISLAND OF REIL

The **insula,** or *island of Reil,* is an island of cortical material located in the depths of the posterior ramus of the lateral fissure (Fig. 4b). It is revealed if the temporal lobe is retracted or if portions of the temporal, frontal, and parietal lobes are removed. The base of the insula is surrounded by the **limiting fissure** (12–4), or *circular fissure*. The **central fissure of the insula** (14–4) divides the structure into anterior and posterior portions. The posterior part is made up of the **long gyrus** (15–4) of the insula, and the anterior part includes the **precentral gyrus of the insula** (13–4) and several short gyri, the **gyri breves** (16–4). The most anterior portion of this buried part of cortex is referred to as the **limen** (17–4) of the insula.

Figure 4

The lobes of the brain and the insula.

(a) Lobes of the brain.

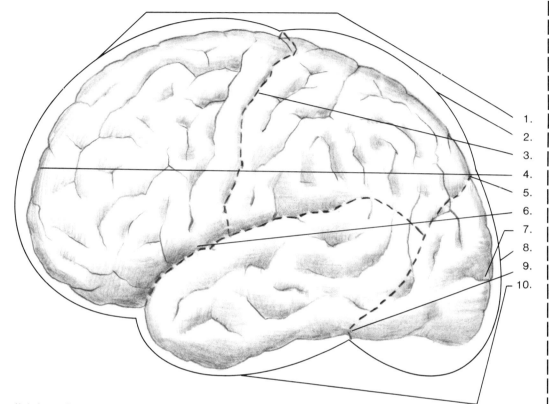

1.	Frontal lobe
2.	Parietal lobe
3.	*Central fissure
4.	Frontal pole
5.	Parieto-occipital fissure
6.	*Lateral fissure
7.	Occipital pole
8.	Occipital lobe
9.	Preoccipital notch
10.	Temporal lobe

(b) Insula.

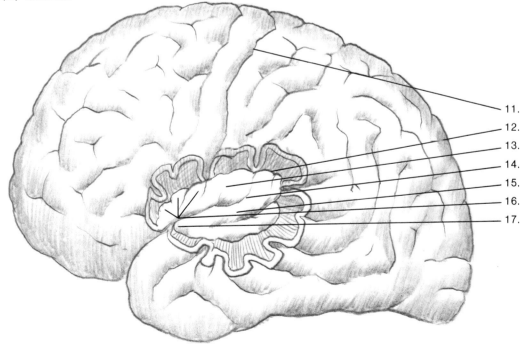

11.	*Central fissure
12.	*Limiting fissure
13.	Precentral gyrus of insula
14.	Central fissure of insula
15.	Long gyrus
16.	Gyri breves
17.	Limen

THE CEREBRAL HEMISPHERES: MEDIAL ASPECT

Figure 5 presents the medial aspect of the cerebral hemisphere: the midsagittal plane. The thalamus, mesencephalon, and rhombencephalon have been removed to expose the parts of the hemisphere that would otherwise be hidden in this section.

The **corpus callosum,** a large commissure connecting the two hemispheres, is one of the most prominent features of the medial surface of the brain. The most rostral portion of this commissure has a sharp bend downward. That portion is referred to as the **genu of the corpus callosum** (2–5). It thins out to form the **rostrum** (21–5). The most posterior portion is thickened and is called the **splenium** (16–5). The main portion of the corpus callosum is the **body** (15–5). The corpus callosum is separated from the cortex above it by the **sulcus of the corpus callosum** (3–5). The **cingulate gyrus** (12–5) lies just dorsal to the corpus callosum and follows the curve of that large commissure. Rostrally the cingulate gyrus curves around the genu and merges with the **subcallosal gyrus** (24–5). Caudally, it curves around the splenium and becomes continuous through the **isthmus** (20–5) with the **hippocampal gyrus** (22–5), which curves back in a dorsal posterior direction to form the **uncus** (27–5). These gyri—the cingulate gyrus, the isthmus, the hippocampal gyrus, and the uncus—form the **limbic lobe,** which is sometimes referred to as the *fornicate lobe.* (See Sect. 41 for more detail on the limbic lobe.)

The cingulate gyrus is bounded dorsally by the **cingulate sulcus** (4–5), which follows the contour of the cingulate gyrus; in the anterior portion it curves ventrally around that gyrus and separates it from the **superior frontal gyrus** (1–5), which lies on both the lateral and the medial surfaces of the hemisphere. The dorsal portion of the cingulate sulcus has two branches: the **paracentral sulcus** (10–5), which is just anterior to the medial extension of the **central fissure** (9–5), and the **marginal sulcus** (8–5), which is just posterior to the central fissure. The paracentral and marginal sulci of the cingulate sulcus form the borders of the **paracentral lobule** (5–5). The anterior portion of this lobule continues dorsally to the lateral surface of the hemisphere and forms the **precentral gyrus** (11–3), while the posterior portion forms the **postcentral gyrus** (6–3) on the lateral surface. That portion of the cingulate sulcus posterior to the marginal branch is referred to as the **subparietal sulcus** (11–5).

The **superior parietal gyrus** (7–5) extends from the lateral aspect of the hemisphere into the medial aspect. Posteriorly it is separated from the occipital lobe by the deep **parieto-occipital sulcus** (13–5). Anteriorly it is bounded by the marginal sulcus. This section of the medial portion of the parietal lobe is called the **precuneus** (6–5).

The area bounded by the parieto-occipital sulcus and the **calcarine fissure** (17,18–5) is part of the occipital lobe and is known as the **cuneus** (14–5). The calcarine fissure extends from the posterior aspect of the occipital lobe anteriorly to a point just below the splenium of the corpus callosum. That portion of the calcarine fissure rostral to the parieto-occipital sulcus is referred to as the **anterior calcarine fissure** (17–5), while that portion caudal to the parieto-occipital sulcus is known as the **posterior calcarine fissure** (18–5). The calcarine fissure also forms the superior border for the second portion of the occipital lobe, the **lingual gyrus** (19–5). Anterior to the lingual gyrus is the **hippocampal gyrus** (22–5). The lingual gyrus is bounded below by the **collateral sulcus** (26–5), which forms the upper boundary of the **fusiform gyrus** (25–5) or *occipitotemporal gyrus.*

Just anterior to the rostrum of the corpus callosum are two cortical fields that are a portion of the **rhinencephalon** (see Fig. 53). The **subcallosal gyrus** (24–5) lies directly against the rostrum and is bounded in the front by the **posterior parolfactory sulcus** (23–5), which is the caudal limit of the **parolfactory area** (28–5). The latter is separated anteriorly from the superior frontal gyrus by the **anterior parolfactory sulcus** (29–5). At the anterior ventral portion of the medial surface is located the **gyrus rectus** (30–5), which can also be seen on the ventral surface of the brain.

Figure 5

Medial surface of the right hemisphere with midbrain and hindbrain removed.

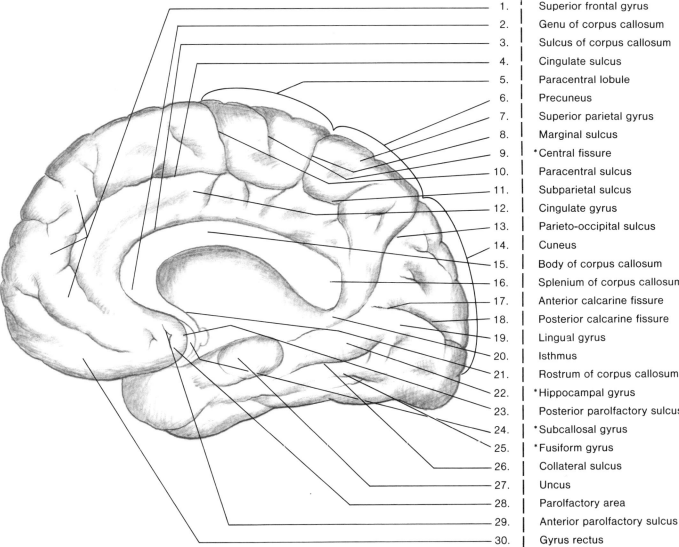

1. Superior frontal gyrus
2. Genu of corpus callosum
3. Sulcus of corpus callosum
4. Cingulate sulcus
5. Paracentral lobule
6. Precuneus
7. Superior parietal gyrus
8. Marginal sulcus
9. *Central fissure
10. Paracentral sulcus
11. Subparietal sulcus
12. Cingulate gyrus
13. Parieto-occipital sulcus
14. Cuneus
15. Body of corpus callosum
16. Splenium of corpus callosum
17. Anterior calcarine fissure
18. Posterior calcarine fissure
19. Lingual gyrus
20. Isthmus
21. Rostrum of corpus callosum
22. *Hippocampal gyrus
23. Posterior parolfactory sulcus
24. *Subcallosal gyrus
25. *Fusiform gyrus
26. Collateral sulcus
27. Uncus
28. Parolfactory area
29. Anterior parolfactory sulcus
30. Gyrus rectus

THE CEREBRAL HEMISPHERES: VENTRAL ASPECT

Figure 6 represents the ventral aspect of the cerebral hemispheres. The brain stem has been removed so that the posterior convolutions may be seen. The ventral surface of the hemisphere reveals two major sections. The larger, posterior section consists of the undersurface of the temporal and occipital lobe, while the anterior portion consists of the orbital section of the frontal lobe.

The most medial convolution of the frontal lobe is the **gyrus rectus** (8–6), which is bordered laterally by the **olfactory sulcus** (2–6). The **orbital gyri** lie lateral to the olfactory sulcus and consist of a number of irregular gyri that vary somewhat from individual to individual. Most frequently the **orbital sulci** (1–6) form a crude letter "H" and are named according to their positions in relation to it, that is, **medial orbital gyrus** (6–6), **anterior orbital gyrus** (3–6), **posterior orbital gyrus** (7–6), and finally **lateral orbital gyrus** (5–6).

The **olfactory bulb (4–6) and olfactory stalk** (9–6) lie directly over the olfactory sulcus, covering all except its most anterior portion. The olfactory stalk proceeds in a posterior direction from the bulb and divides into two striae, the **lateral olfactory stria** and the **medial olfactory stria** (11–6). The bifurcation of these striae enclose a triangular area known as the **olfactory trigone** (12–6). Just posterior to it is the **anterior perforated substance** (13–6), so named because the area is penetrated by numerous blood vessels that penetrate the brain and supply the depths of the hemisphere. The anterior perforated substance is bounded posteriorly by the **optic tract** (14–6).

The rest of the ventral aspect of the hemisphere is essentially occupied by the temporal lobe. The ventral surface of the temporal lobe is concave and extends from the **temporal pole** (10–6) caudally to the **occipital pole** (24–6) and is continuous with the ventral surface of the occipital lobe. The anterior tip of the temporal lobe overlaps the more posterior portion of the orbital gyri. The **inferior temporal gyrus** (15–6) constitutes the most lateral portion of the ventral aspect of the temporal lobe and is also represented on the lateral surface of the hemisphere. It is bounded medially by the **inferior temporal sulcus** (21–6), which runs from near the temporal pole to near the occipital pole. This sulcus separates the inferior temporal gyrus from the **fusiform gyrus** (19–6), or *occipitotemporal gyrus.* The most posterior portions of the inferior temporal gyrus and the fusiform gyrus are sometimes considered to be a part of the occipital lobe.

The **collateral fissure** (23–6) separates the fusiform gyrus from the **hippocampal gyrus** (20–6), or *parahippocampal gyrus,* the anterior portion of which curves back on itself to form the **uncus** (16–6). The most posterior portion of the uncus is referred to as the **intralimbic gyrus** (18–6). The hippocampal gyrus is bounded above by the **hippocampal fissure** (17–6). Some anatomists consider the hippocampal gyrus and uncus to be a part of the temporal lobe, while others suggest they form a part of the **rhinencephalic cortex**. The **lingual gyrus** (22–6) can be seen as the posterior extension of the hippocampal gyrus.

Figure 6

Ventral aspect of the cerebral hemispheres with brain stem removed.

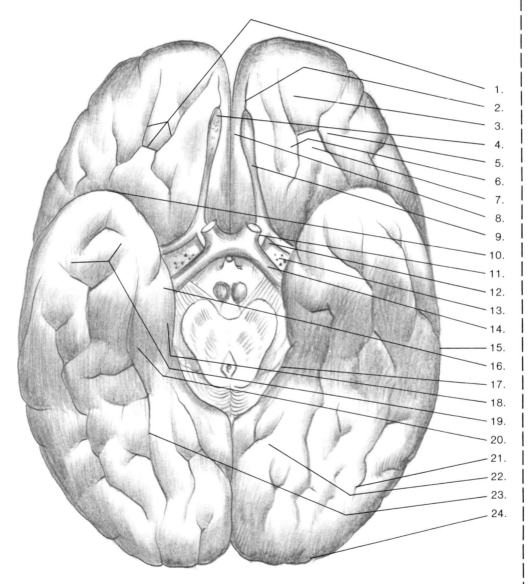

1.   Orbital sulci
2.   Olfactory sulcus
3.   Anterior orbital gyrus
4.   Olfactory bulb
5.   Lateral orbital gyrus
6.   Medial orbital gyrus
7.   Posterior orbital gyrus
8.   Gyrus rectus
9.   Olfactory stalk
10. Temporal pole
11. Medial olfactory stria
12. Olfactory trigone
13. Anterior perforated substance
14. Optic tract
15. Inferior temporal gyrus
16. Uncus
17. Hippocampal fissure
18. Intralimbic gyrus
19. *Fusiform gyrus
20. *Hippocampal gyrus
21. Inferior temporal sulcus
22. Lingual gyrus
23. Collateral fissure
24. Occipital pole

THE BRAIN STEM: LATERAL ASPECT

Figures 7–9 provide a surface view of the brain stem and of its many landmarks that will serve as reference points throughout the study of neuroanatomy. The lateral and posterior surfaces of the brain stem are normally hidden from view by the cerebellum and the cerebral hemispheres. In this series of diagrams, those structures have been removed so that the entire surface of the brain stem can be studied. Figure 7 presents a lateral view of the brain stem.

The **medulla oblongata** (33–7) is continuous with the spinal cord and extends from the point of the most rostral rootlet of the **first cervical nerve** (31–7) to the caudal border of the **pons** (32–3). It is shaped like a truncated cone with its smallest diameter (about 2 centimeters) at the junction with the spinal cord.

The lateral aspect of the medulla shows two sulci that are continuous with those on the lateral surface of the spinal cord (8–14), (1–15). The **ventrolateral sulcus** (29–7) runs the length of the medulla, and at its more rostral extent the fibers of the **abducens nerve [VI]** (19–7) emerge from it along with the fibers of the **hypoglossal nerve [XII]** (28–7) just below them. The **dorsolateral sulcus** (23–7) is the groove from which the **glossopharyngeal nerve [IX]** (21–7), the **vagus nerve [X]** (24–7), and the **accessory nerve [XI]** (30–7) emerge.

A prominent ovoid body, the **inferior olive** (25–7), lies between the anterior lateral sulcus and the posterior lateral sulcus at the level of emergence of the hypoglossal nerve [XII] and the glossopharyngeal [IX] and vagus [X] nerves. This ovoid landmark is caused by a mass of gray substance, the **inferior olivary nucleus** (25–7), which is located just below the surface. A number of fine fiber bundles can be seen on the surface of the olive. They originate in the **ventral median fissure** and in the ventrolateral sulcus and run dorsally over the olive and the nearby medullary surface to join the **inferior cerebellar peduncle** (20–7), or *restiform body*. These are the **ventral external arcuate fibers** (27–7). Just posterior to the dorsolateral sulcus at the level of the emergence of the roots of the glossopharyngeal [IX], vagus [X], and accessory [XI] nerves lies a club-shaped eminence called the **tuberculum cinereum** (26–7). This body is formed by descending fibers from the sensory root of the **trigeminal nerve [V]** (14–7) and by the one of the nuclei of that nerve, the **substantia gelatinosa** (9–16).

The **acoustic nerve [VIII]** (17–7), also called the *vestibulocochlear nerve* or *statoacoustic nerve*, lies in a depression between the rostral portion of the olive and the pons. Just posterior to the acoustic nerve is another slight depression from which the **facial nerve [VII]** (18–7) emerges.

The pons, which constitutes the ventral portion of the metencephalon, is continuous with the medulla oblongata and is separated from it by a shallow furrow, the **inferior pontine sulcus** (22–7). It is essentially a mass of transverse fibers that extend from pontine nuclei into the two halves of the cerebellum. These fibers constitute the principal portion of the **middle cerebellar peduncles** (16–7), or *brachium pontis*. These transverse fibers connect the cerebral hemispheres with the opposite cerebellar hemispheres. The two roots of the **trigeminal nerve [V]** (14–7) exit from the lateral aspect of the pons at the level of the middle cerebellar peduncle. The larger portion of the trigeminal nerve is sensory in function and is called the **portio major** (15–7). Just rostral to that is the **portio minor** (13–7), which has a motor function. The inferior cerebellar peduncle, or restiform body, connects the cerebellum and the medulla. In Figure 7 it can be seen just below the pons at the most posterior portion.

The **midbrain** (12–7), or **mesencephalon,** lies between the pons and the diencephalon and is separated from the former by the **superior pontine sulcus** (7–7). The most prominent feature of the lateral aspect of the midbrain is the massive **cerebral peduncle** (6–7), or *basis pedunculi* also called the *crus cerebri*. This large bundle is primarily composed of motor fibers descending from the cerebrum to the more caudal portions of the nervous system. At about the point where the peduncle emerges from the cerebrum the **optic tract** (32–7) curves around its ventrolateral surface. The **trochlear nerve [IV]** (10–7) curves around the surface of the peduncle just above the superior pontine sulcus. On the more dorsal portion of the lateral aspect of the midbrain the **medial geniculate bodies** (8–7), the **superior colliculi** (9–7), and the **inferior colliculi** (11–7) can be seen.

If the cerebral peduncles are followed rostrally, they fan out into a very broad band of fibers that enter the cerebral hemispheres. Most of the fibers turn in a medial direction and form the **internal capsule** (3–8). A smaller number are deflected laterally to form the **external capsule** (2–8).

The internal and external capsules enclose the **lenticular nucleus** (5–8). The two capsules come together above the lenticular nucleus to form a major radiation

(Continued)

Figure 7

Lateral view of the brain stem.

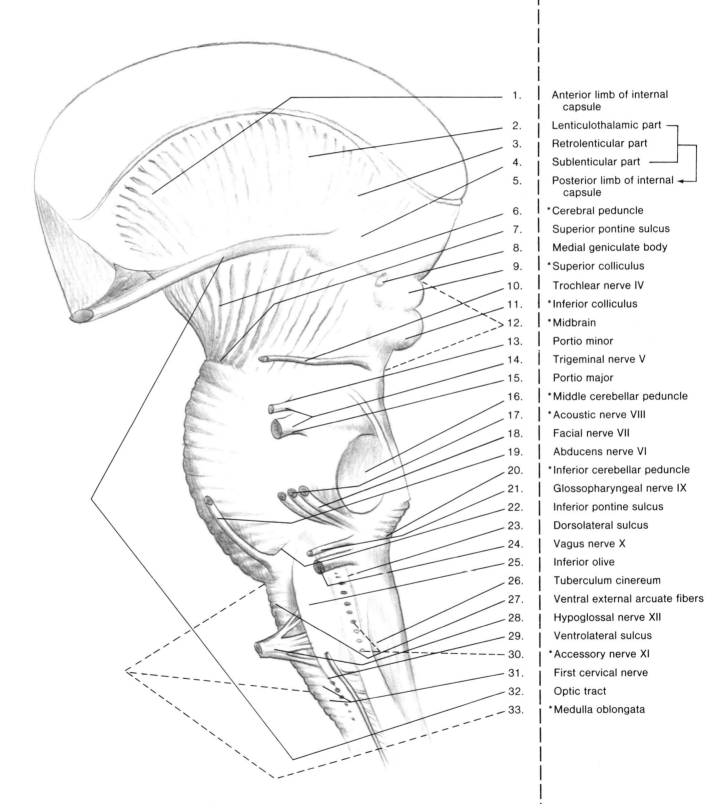

1. Anterior limb of internal capsule
2. Lenticulothalamic part
3. Retrolenticular part
4. Sublenticular part
5. Posterior limb of internal capsule
6. *Cerebral peduncle
7. Superior pontine sulcus
8. Medial geniculate body
9. *Superior colliculus
10. Trochlear nerve IV
11. *Inferior colliculus
12. *Midbrain
13. Portio minor
14. Trigeminal nerve V
15. Portio major
16. *Middle cerebellar peduncle
17. *Acoustic nerve VIII
18. Facial nerve VII
19. Abducens nerve VI
20. *Inferior cerebellar peduncle
21. Glossopharyngeal nerve IX
22. Inferior pontine sulcus
23. Dorsolateral sulcus
24. Vagus nerve X
25. Inferior olive
26. Tuberculum cinereum
27. Ventral external arcuate fibers
28. Hypoglossal nerve XII
29. Ventrolateral sulcus
30. *Accessory nerve XI
31. First cervical nerve
32. Optic tract
33. *Medulla oblongata

known as the **corona radiata** (1–8). In Figure 7 the external capsule and lenticular nucleus have been removed to show the branches of the internal capsule. The internal capsule is divided into two major limbs, the **anterior limb of the internal capsule** (1–7) and the **posterior limb of the internal capsule** (5–7). The latter is further subdivided into the **lenticulothalamic part** (2–7), the **retrolenticular part** (3–7), and the **sublenticular part** (4–7). The relationship between the lenticular nuclei, the internal and external capsules, and the thalamus and the caudate nucleus will become clear in later figures.

THE BRAIN STEM: DORSAL ASPECT

Figure 8 shows the dorsal view of the brain stem with the cerebellum cut away to reveal the structures beneath it. The three large bands of fibers leading to the cerebellum have been cut; they appear in the figure as white areas. The most caudal cerebellar peduncle, the **inferior cerebellar peduncle** (40–8), or *restiform body*, extends from the medulla in a dorsal direction into the cerebellum. The **middle cerebellar peduncle** (35–8), or *brachium pontis*, extends from the pons into the cerebellum, while the **superior cerebellar peduncle** (30–8), or *brachium conjunctivum*, runs from the anterior cerebellum into the area of the brain stem just below the midbrain.

The dorsal aspect of the medulla shows the sulci, which are a continuation of those in the spinal cord. The **dorsal median sulcus** (56–8), as the name implies, runs longitudinally in the center of the medulla, separating it into two halves. Just lateral to this sulcus is the **dorsal intermediate sulcus** (57–8). The **gracilis tubercle** (54–8), also referred to as the *clava*, lies between the aforementioned sulci, and the **cuneate tubercle** (52–8) lies just lateral to the dorsal intermediate sulcus. These two eminences are formed by the **nucleus cuneatus** (7–21) and the **nucleus gracilis** (5–21) just below the surface.

The central canal continues from the spinal cord into the medulla. At about the middle of the medulla the canal widens into the **fourth ventricle** (36–8), which extends from the central canal to the **cerebral aqueduct** (20–10), or *aqueduct of Sylvius*, in the midbrain. The roof of the fourth ventricle is formed caudally by the thin **tela choroidea** (44–8), which passes into the **posterior medullary velum** (48–8). Rostral to the posterior medullary velum, the roof of the ventricle is formed by the white matter of the cerebellum. This white matter is continuous with the **anterior medullary velum** (26–8), which covers the portion of the ventricle in the area of the **pons** (32–3). The **obex** (53–8) is the small triangular plate at the most inferior point of the tela choroidea.

When the cerebellum and the **anterior medullary velum** (26–8) and **posterior medullary velum** (48–8) are cut away, the floor of the fourth ventricle is exposed. It is referred to as the **rhomboid fossa** (33–8) and is divided into two symmetric lateral halves by the **median sulcus** (31–8). The **sulcus limitans** (39–8) lies lateral to the median sulcus. The superior and inferior portions of this sulcus are deeper than the middle portion and are named the **superior fovea** (38–8) and the **inferior fovea** (47–8), respectively. Several strands of fibers known as the **striae cerebellares** (41–8) cross the rhomboid fossa in its intermediate portion. It is generally agreed that they run into the cerebellum.

The sulcus limitans divides each half of the rhomboid fossa into two sections. The medial portion is the **median eminence** (34–8), while the more lateral, triangular portion is the **vestibular area** (42–8), which covers the **nuclei of the vestibular nerve [VIII]** (17–65). Just lateral to the vestibular area is the **acoustic tubercle** (43–8), which covers the **nucleus of the acoustic nerve [VIII]** (18–65). The caudal portion of the median eminence covers the underlying nucleus of the **hypoglossal nerve [XII]** (24–65) and is known as the **trigonum hypoglossi** (45–8). More rostrally the median eminence broadens out into a rounded eminence, the **facial colliculus** (37–8). Just lateral to the trigonum hypoglossi is the **trigonum vagi** (46–8), also called the *ala cinerea*, which covers the dorsal nuclei of the **vagus nerve [X]** (23–65). Just lateral to the trigonum vagi is the **funiculus separans** (49–8), a ridge of tissue that separates the trigonum vagi from a narrow zone that borders the lateral wall of the ventricle, the **area postrema** (51–8), the function of which is unknown.

The **locus ceruleus** (32–8) is a shallow groove, usually blue in color, which extends from the **superior fovea** (38–8) rostrally to the cerebral aqueduct. The most inferior section of the rhomboid fossa resembles an old-fashioned pen and was called the **calamus scriptorius** (50–8) by early anatomists.

The dorsal aspect of the mesencephalon is called the **tectum** (23–8), or *quadrigeminal plate*, also referred to as *corpora quadrigemina* or *quadrigeminal bodies*. The most prominent features on the tectum are four eminences, the **superior colliculi** (16–8) and the **inferior colliculi** (19–8). The superior colliculi are associated with visual processes, and the inferior colliculi are involved in hearing. The inferior colliculi are hemispherical in shape, while the superior tend to be more oval and are somewhat darker in color. The colliculi are separated by a longitudinal groove that terminates rostrally in the **pineal body** (15–8) and caudally in the **frenulum veli** (25–8), which is continuous with the anterior medullary velum. The anterior medullary velum, just below the frenulum veli, is covered with a thin lobule of the cerebellum, the **lingula** (29–8).

An arm, or brachium, extends from each of the colliculi in a lateral direction to connect them with the **genic-**

(Continued)

Figure 8

Dorsal view of the brain stem.

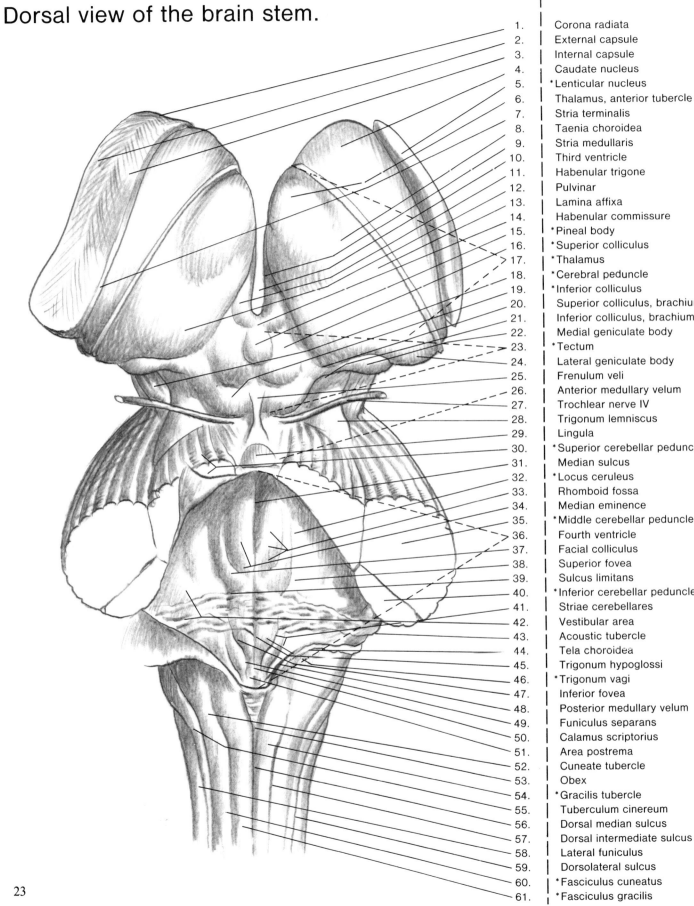

1. Corona radiata
2. External capsule
3. Internal capsule
4. Caudate nucleus
5. *Lenticular nucleus
6. Thalamus, anterior tubercle
7. Stria terminalis
8. Taenia choroidea
9. Stria medullaris
10. Third ventricle
11. Habenular trigone
12. Pulvinar
13. Lamina affixa
14. Habenular commissure
15. *Pineal body
16. *Superior colliculus
17. *Thalamus
18. *Cerebral peduncle
19. *Inferior colliculus
20. Superior colliculus, brachium
21. Inferior colliculus, brachium
22. Medial geniculate body
23. *Tectum
24. Lateral geniculate body
25. Frenulum veli
26. Anterior medullary velum
27. Trochlear nerve IV
28. Trigonum lemniscus
29. Lingula
30. *Superior cerebellar peduncle
31. Median sulcus
32. *Locus ceruleus
33. Rhomboid fossa
34. Median eminence
35. *Middle cerebellar peduncle
36. Fourth ventricle
37. Facial colliculus
38. Superior fovea
39. Sulcus limitans
40. *Inferior cerebellar peduncle
41. Striae cerebellares
42. Vestibular area
43. Acoustic tubercle
44. Tela choroidea
45. Trigonum hypoglossi
46. *Trigonum vagi
47. Inferior fovea
48. Posterior medullary velum
49. Funiculus separans
50. Calamus scriptorius
51. Area postrema
52. Cuneate tubercle
53. Obex
54. *Gracilis tubercle
55. Tuberculum cinereum
56. Dorsal median sulcus
57. Dorsal intermediate sulcus
58. Lateral funiculus
59. Dorsolateral sulcus
60. *Fasciculus cuneatus
61. *Fasciculus gracilis

ulate bodies (22,24–8). The **brachium of the inferior colliculus** (21–8), a short, flat band of fibers extends from the inferior colliculus to the **medial geniculate body** (22–8). The **brachium of the superior colliculus** (20–8) extends from the superior colliculus to the **lateral geniculate body** (24–8). Some of the fibers of this tract continue into the **optic tract** (32–7). Just caudal to the inferior colliculi the **trochlear nerve [IV]** (27–8) emerges and runs laterally over a small triangular section of the midbrain known as the **trigonum lemnisci** (28–8), which is separated from the **cerebral peduncle** (18–8) by the **lateral mesencephalic sulcus.** The trigonum lemnisci is composed primarily of fibers from the **lateral lemniscus** (11–24).

The prominent ovoid mass rostral to the superior colliculi is the **thalamus** (17–8), a part of the diencephalon. Laterally, it is separated from the **caudate nucleus** (4–8) by a slender bundle of fibers, the **stria terminalis** (7–8). The body of the thalamus in the dorsal view can be divided into a medial and a lateral portion by an oblique groove, the **taenia choroidea** (8–8). The thin covering of the area lateral to that groove is called the **lamina affixa** (13–8). The anterior portion of the medial part is the **anterior tubercle of the thalamus** (6–8), and the posterior part is the **pulvinar** (12–8). The **third ventricle** (10–8) lies between the two thalami. A thin strip of tissue, the **stria medullaris** (9–8), separates the medial and dorsal surfaces of the thalamus. Caudally it becomes more broad and forms the **habenular trigone** (11–8), the two halves of which are joined by the **habenular commissure** (14–8).

Anterolateral to the stria terminalis lies the **caudate nucleus** (4–8), bounded laterally by the **internal capsule** (3–8), which with the **external capsule** (2–8) forms the **corona radiata** (1–8). On the right-hand side of the figure the internal and external capsules have been deleted to show the relative position of the **lenticular nucleus** (5–8).

THE BRAIN STEM: VENTRAL ASPECT

The ventral aspect of the brain stem shows many of the features already discussed in the section on the lateral aspect. Figure 9 provides a different view and will facilitate learning the important landmarks.

The **ventral median sulcus** (39–9) in the brain stem is continuous with that in the spinal cord. It continues rostrally to end at the lower border of the pons, where it expands into a triangular shape, the **foramen cecum** (25–9). The **ventrolateral sulcus** (35–9) also continues from the spinal cord to the inferior border of the pons, the **inferior pontine sulcus** (24–9). Between these two sulci lie the most prominent features of the ventral medulla, the **pyramids** (22–20). These two large bundles of fibers contain the motor nerves that run from the brain through the brain stem to the spinal cord. About two-thirds of these fibers leave the pyramid and cross the ventral median sulcus to the opposite side, forming an easily recognized landmark, the **pyramidal decussation** (37–9). The remaining fibers do not cross at the pyramidal decussation but continue down the lateral portion of the pyramids into the spinal cord.

The **dorsolateral sulcus** (36–9) can also be seen in this figure, and as indicated earlier the rootlets of the **accessory nerve [XI]** (34–9) emerge from that sulcus and proceed in a rostral direction. The **inferior olive** (33–9) is again shown with the **hypoglossal nerve [XII]** (32–9) emerging from the ventrolateral sulcus between the olive and the pyramid. At about the same level, the **vagus nerve [X]** (31–9) and the **glossopharyngeal nerve [IX]** (30–9) emerge from the dorsolateral sulcus.

Several of the cranial nerves exit from the point of juncture between the pons and the medulla. The most medial is the **abducens nerve [VI]** (22–9). Lateral to it is the **facial nerve [VII]** (26–9) with its sensory branch, the **intermediate nerve [VII]** (27–9). Even more laterally is the **acoustic nerve [VIII]** (23–9), also called the *vestibulocochlear nerve*. As can be seen in Figure 9 the **vestibular** (28–9) and the **acoustic** (29–9) portions of this eighth cranial nerve separate before they enter the brain stem, the acoustic division being lateral to the vestibular.

The **pons** (39–10) is convex in shape and has a medial furrow, the **basilar groove** (16–9), in which lies the basilar artery. On either side of the basilar groove there is a bulge, or eminence, caused by the underlying traverse of the cerebrospinal fibers through the body of the pons. The **trigeminal nerve [V]** (21–9) with its two divisions, sensory, **portio minor** (19–9), and motor, **portio major** (20–9), also emerges from the lateral pons. A longitudinal line drawn through the trigeminal nerve is generally taken as the division between the ventral surface of the pons and the **middle cerebellar peduncle** (18–9).

The **superior pontine sulcus** (13–9) is the rostral border of the pons. Just above it the large **cerebral peduncle** (12–9) can be seen running in a rostral, slightly lateral direction. The **trochlear nerve [IV]** (17–9) comes from its dorsal exit around the peduncles and is visible from the ventral view. Just above the pons at the midline the **oculomotor nerves [III]** (11–9) exit. Between the peduncles is an indentation known as the **interpeduncular fossa** (10–9). This gray substance is pierced by a number of blood vessels and because of this is also referred to as the *posterior perforated substance*.

The **mammillary bodies** (9–9) lie just rostral to the interpeduncular fossa. They are hemispherical white bodies about the size of small peas. The **tuber cinereum** (7–9) lies just rostral to them. The **infundibulum** (6–9) projects downward from the tuber cinereum to join the **pituitary** (38–10), or *hypophysis*.

The **optic chiasm** (5–9) is the junction between the **optic nerves** [II] (4–9) and the **optic tracts** (8–9). The optic nerves originate in the retina and undergo a partial decussation in the chiasm. The optic tracts travel in a latero-caudal direction across the cerebral peduncles and divide into a **lateral root** (14–9) and a **medial root** (15–9). Most of the fibers of the lateral root end up in the lateral geniculate body.

The brain has been dissected away above the optic tract to help clarify the relationship between the **lenticular nucleus** (3–9), the **caudate nucleus** (1–9), and the **internal capsule** (2–9).

Figure 9

Ventral view of the brain stem.

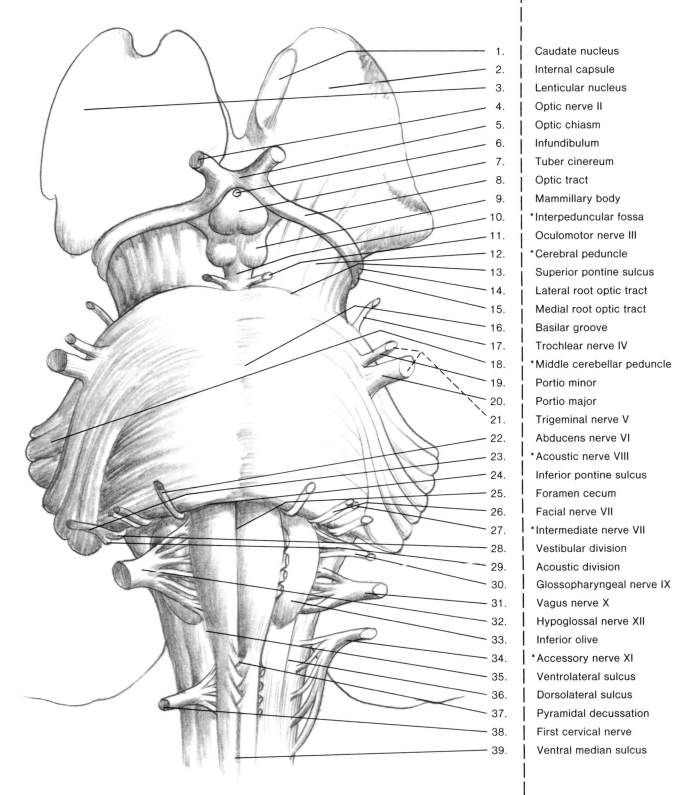

1.   Caudate nucleus
2.   Internal capsule
3.   Lenticular nucleus
4.   Optic nerve II
5.   Optic chiasm
6.   Infundibulum
7.   Tuber cinereum
8.   Optic tract
9.   Mammillary body
10. *Interpeduncular fossa
11. Oculomotor nerve III
12. *Cerebral peduncle
13. Superior pontine sulcus
14. Lateral root optic tract
15. Medial root optic tract
16. Basilar groove
17. Trochlear nerve IV
18. *Middle cerebellar peduncle
19. Portio minor
20. Portio major
21. Trigeminal nerve V
22. Abducens nerve VI
23. *Acoustic nerve VIII
24. Inferior pontine sulcus
25. Foramen cecum
26. Facial nerve VII
27. *Intermediate nerve VII
28. Vestibular division
29. Acoustic division
30. Glossopharyngeal nerve IX
31. Vagus nerve X
32. Hypoglossal nerve XII
33. Inferior olive
34. *Accessory nerve XI
35. Ventrolateral sulcus
36. Dorsolateral sulcus
37. Pyramidal decussation
38. First cervical nerve
39. Ventral median sulcus

THE BRAIN STEM: SAGITTAL ASPECT

The view of the brain stem in Figure 10 completes the perspective and shows the relationship among the various landmarks from a different aspect. Since it is a sagittal view, some internal structures that were hidden in the earlier figures can be seen.

This midline view of the brain stem shows the continuity of the ventricular system within it. It can be seen that the **central canal** (40–10) of the spinal cord opens up into the **fourth ventricle** (27–10), which flows into the narrowed section of the midbrain called the **cerebral aqueduct** (20–10), or *aqueduct of Sylvius.* That narrow canal opens rostrally into the **third ventricle** (13–10).

The medulla and pons form the ventral surface of the fourth ventricle, the roof of which is peaked. That peaked portion is referred to as the **fastigium** (25–10). For a short distance caudally from the fastigium the roof of the fourth ventricle is composed of the **posterior medullary velum** (28–10). It becomes continuous with the **tela choroidea** (33–10). The tela choroidea has projecting from it highly vascularized loops of tissue covered by a modified glandular ependymal epithelium that make up the **choroid plexus** (34–10), which projects into the fourth ventricle at that point. Just caudal to the choroid plexus is an opening in the roof of the fourth ventricle that communicates with the **subarachnoid space** (36–10). It is called the **medial aperture of the fourth ventricle** (35–10), or *foramen of Magendie.* Two other apertures not shown on this figure occur at the lateral extensions of the ventricle and are called **lateral apertures,** or *foramen of Luschka.*

Rostral to the fastigium the pontile portion of the ventricle is roofed by the **anterior medullary velum** (26–10), which extends to the midbrain and becomes continuous with the **tectum** (18–10). The **lingula** (24–10) is the thickened portion of the anterior medullary velum.

The tectum, composed of the **inferior colliculi** (19–10) and **superior colliculi** (15–10), merges with a significant bundle of fibers, the **posterior commissure** (14–10). Some of these fibers connect the two superior colliculi, but the origin and destination of most of them are, as yet, obscure.

In cross section the **pineal body** (12–10) is shaped like the cone of a fir tree and lies upon the anterior mesencephalon, to which it is connected by the **habenular commissure** (11–10). The **habenula** (10–10) is located just anterior to its commissure and receives fibers from the **stria medullaris** (7–10), which arches over the interior lateral surfce of the **third ventricle** (13–10). The wall of the third ventricle is the medial aspect of the **thalamus** (6–10). The anterior and dorsal boundaries of the third ventricle form the **fornix** (2,5–10), which, although it cannot be seen in this figure, is connected to the **mammillary body** (23–10) and curves dorsally to enter the cerebral hemisphere and connect with the hippocampus. In this figure it emerges just behind the **anterior commissure** (17–10) and disappears under the **corpus callosum** (1–10). The dorsal section of the fornix is called the **body** (2–10); the anterior portion is the **column** (5–10).

Between the fornix and the corpus callosum stretches a thin membranous plate, the **septum pellucidum** (3–10). It separates the two lateral ventricles of the cerebral hemispheres. The **choroid plexus of the third ventricle** (4–10) can be seen extending over the entire roof of the third ventricle from the **interventricular foramen** (8–10) caudally to the habenular complex. Like the choroid plexus of the fourth ventricle, it is composed of highly vascularized tissue. The interventricular foramen is a communication channel between the third and the lateral ventricles. (For a complete picture of the ventricles of the brain and of the circulation of the cerebrospinal fluid, see Figs. 61 and 62.)

In the center of the thalamus is the **intermediate mass** (9–10), also called the *interthalamic adhesion* or *massa intermedia.* This is a concentration of gray matter connecting the two thalami in about 70 percent of human brains.

The **anterior commissure** (17–10) is a significant band of fibers connecting a variety of structures in the two hemispheres. It is just above the **lamina terminalis** (22–10), which is also a commissure. The **optic chiasm** (32–10), seen earlier in the ventral aspect, is shown here in cross section. Just below it, an indentation, the **optic recess** (30–10), can be seen. The **infundibulum** (37–10) emerges from the **tuber cinereum** (31–10) to form the stalk of the **pituitary body** (38–10), or *hypophysis.*

The mammillary body lies caudal to the tuber cinereum. Both the mammillary body and tuber cinereum are a part of the **hypothalamus** (15–29), which is part of the diencephalon. The hypothalamus is bounded dorsally by the **hypothalamic sulcus** (16–10) and anteriorly by the anterior commissure, the lamina terminalis, and the optic chiasm.

The **pons** (39–10) is readily seen in cross section. The area of the pons dorsal to the large fiber band of the cerebral peduncle is called the **tegmentum** (29–10) and is continuous with the **tegmentum of the midbrain** (21–10).

Figure 10

Sagittal view of the brain stem.

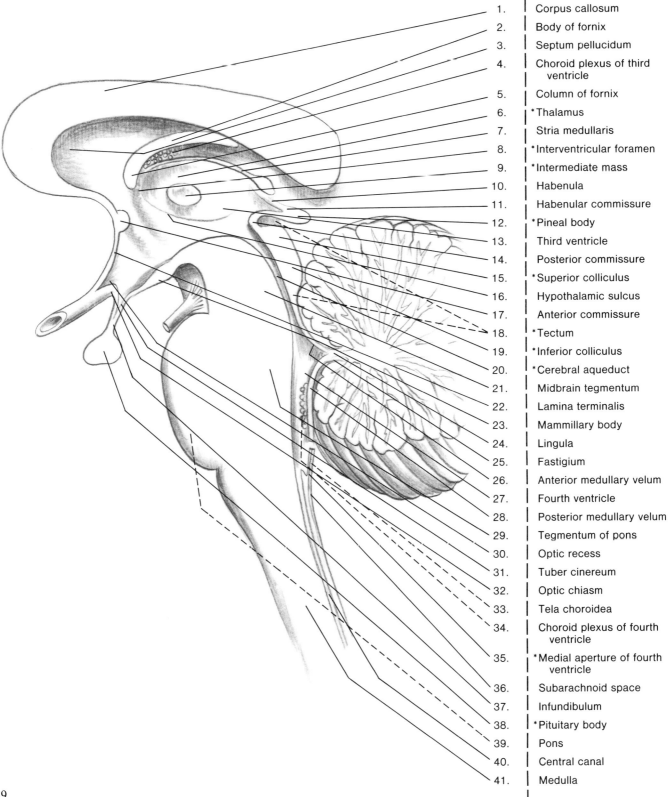

1. Corpus callosum
2. Body of fornix
3. Septum pellucidum
4. Choroid plexus of third ventricle
5. Column of fornix
6. *Thalamus
7. Stria medullaris
8. *Interventricular foramen
9. *Intermediate mass
10. Habenula
11. Habenular commissure
12. *Pineal body
13. Third ventricle
14. Posterior commissure
15. *Superior colliculus
16. Hypothalamic sulcus
17. Anterior commissure
18. *Tectum
19. *Inferior colliculus
20. *Cerebral aqueduct
21. Midbrain tegmentum
22. Lamina terminalis
23. Mammillary body
24. Lingula
25. Fastigium
26. Anterior medullary velum
27. Fourth ventricle
28. Posterior medullary velum
29. Tegmentum of pons
30. Optic recess
31. Tuber cinereum
32. Optic chiasm
33. Tela choroidea
34. Choroid plexus of fourth ventricle
35. *Medial aperture of fourth ventricle
36. Subarachnoid space
37. Infundibulum
38. *Pituitary body
39. Pons
40. Central canal
41. Medulla

THE CEREBELLUM: EXTERNAL ANATOMY

Figures 11, 12, and 13 show the external anatomy of the cerebellum. The figures will be discussed together and therefore should be studied together.

The cerebellum encloses the roof of the fourth ventricle and is, thus, above the medulla, the pons, and the caudal portion of the midbrain. Its dorsal aspect is covered by the cerebral hemispheres, and it is separated from them by that portion of the dura mater (the outermost of the cerebral meninges) known as the **tentorium cerebelli** (13–60).

The cerebellum is functionally important in the coordination of motor responses from the reflex level to the conscious level. Because of its multiple connections with the vestibular system, it also has an important function in the maintenance of equilibrium.

Like the cerebral hemispheres, the cerebellum has an external covering of gray matter that encloses a mass of white. The latter is called the **corpus medullare** (17–11). The folds and invaginations of the cerebellar cortex over the white matter produce a pattern that when seen in cross section resembles a tree and was referred to as the tree of life, **arbor vitae** (12–11), by early anatomists.

Structurally the cerebellum is composed of a central portion, the vermis, which, as its name implies, curves in a wormlike manner around the superior and inferior surfaces. As noted below, each portion of the vermis is given a separate name. The vermis connects the two lateral lobes or hemispheres. As can be seen in Figures 11 and 12, the cortex of the cerebellum is folded into long, slender convolutions called **folia** (28–11), (8–12), which are separated by relatively shallow, parallel sulci. There are several major fissures that are almost deep enough to reach the roof of the fourth ventricle. These fissures divide the cerebellum into lobes and lobules. On the superior surface there is no clear demarcation between the vermis and the two cerebellar hemispheres. Ventrally, however, the vermis is recessed fairly deeply between the hemispheres and appears to be distinct from them. The resultant concave ventral surface of the cerebellum is referred to as the *vallecula cerebelli*, in which the medulla oblongata is situated. The indentation at the point of the vermis in the anterior cerebellum is the **anterior incisure** (17–12). The posterior indentation is the **posterior incisure** (40–12).

The nomenclature for the cerebellum is confused and contradictory, varying considerably from one anatomist to another. Early students of anatomy assigned names on the basis of shape, not function, and like those who take Rorschach inkblot tests each saw something different in the shapes. The terminology most frequently employed is used here.

The lobes consist of the central portion, formed by a segment of the vermis, and two lateral portions, which extend into the cerebellar hemispheres. The **primary fissure** (3–11), (6–12), (6–13), or *preclival fissure,* divides the cerebellum into the **anterior lobe** (1–11), (1–12), (28–13), and **posterior lobe** (31–11), (16–12), (27–13). Both are further divided into lobules. The portion of the lobule formed by the vermis is given a different name from the hemispherical portions. This is made clear by the schematic in Figure 13. The lobule anterior to the primary fissure is the **anterior semilunar lobule** (8–11), (20–12), (5–13), or *anterior crescentic lobule.* The vermis portion of this lobule is called the **culmen** (4–11), (18–12), (7–13). In some texts this lobule is also referred to as the *anterior quadrangular lobule.* The **postcentral fissure** (6–11), (3–12), (4–13) separates the anterior semilunar lobule from the **central lobule** (9–11), (19–12), (3–13), which is separated from the **lingula** (13–11), (25–12), (1–13) by the **precentral fissure** (10–11), (21–12), (2–13).

The most anterior lobule between the primary fissure and the **posterior superior fissure** (5–11), (10–12), (9–13), also called the *postclival fissure,* is the **posterior semilunar lobule** (11–11), (27–12), (8–13), also called the *simple lobe,* the *lobulus simplex,* or the *posterior quadrangular lobule.* The section of the vermis included in this lobe is called the **declive** (2–11), (9–12), (11–13). Between the posterior superior fissure and the **horizontal fissure** (7–11), (14–12), (29–12), (12–13) lies the **superior semilunar lobule** (14–11), (30–12), (10–13), also called *crus I.* The part of the vermis included in this lobule is the **folium** (16–11), (12–12), (13–13). The **inferior semilunar lobule** (24–11), (15–12), (14–13) is also called *crus II.* **Tuber** (21–11), (39–12), (15–13) is the name given to the vermis at this point. Crus I and crus II are also called the *ansiform lobule.* The posterior portion of the inferior semilunar lobule is called the **gracile lobule** (17–13).

The **postpyramidal fissure** (23–11), (34–12), (18–13) lies completely in the ventral aspect of the cerebellum and separates the inferior semilunar lobule from the **biventer lobule** (29–11), (31–12), (19–13), also called *dorsal paraflocculus.* The **pyramis** (26–11), (38–12), (16–13) is that portion of the vermis between the two

(Continued)

Figure 11

Sagittal section through the vermis of the cerebellum showing the fissures and lobules.

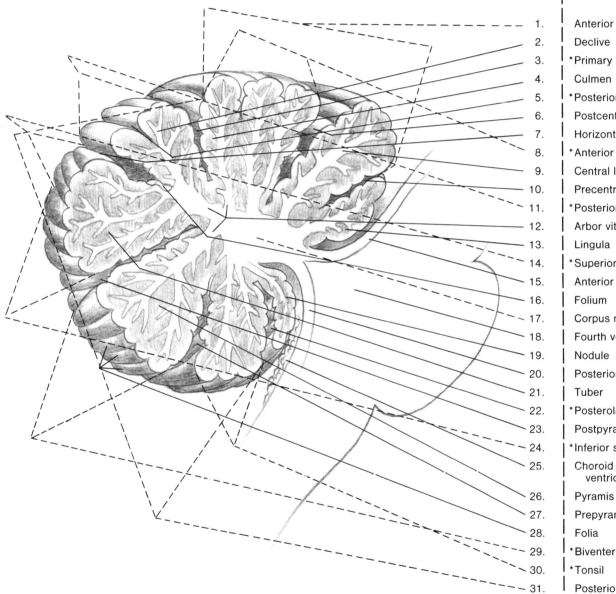

1. Anterior lobe
2. Declive
3. *Primary fissure
4. Culmen
5. *Posterior superior fissure
6. Postcentral fissure
7. Horizontal fissure
8. *Anterior semilunar lobule
9. Central lobule
10. Precentral fissure
11. *Posterior semilunar lobule
12. Arbor vitae
13. Lingula
14. *Superior semilunar lobule
15. Anterior medullary velum
16. Folium
17. Corpus medullare
18. Fourth ventricle
19. Nodule
20. Posterior medullary velum
21. Tuber
22. *Posterolateral fissure
23. Postpyramidal fissure
24. *Inferior semilunar lobule
25. Choroid plexus of fourth ventricle
26. Pyramis
27. Prepyramidal fissure
28. Folia
29. *Biventer lobule
30. *Tonsil
31. Posterior lobe

halves of the biventer lobule. The **prepyramidal fissure** (27–11), (36–12), (20–13) lies between the biventer and a lobule called the **tonsil** (30–11), (33–12), (21–13), or *ventral paraflocculus.* The two tonsils are separated by the portion of the vermis called the **uvula** (35–12), (22–13).

The **posterolateral fissure** (22–11), (32–12), (23–13), also called the *postnodular fissure,* forms the posterior boundary of the posterior lobe and separates it from the **flocculonodular lobe** (26–13), which is made up of the **nodule** (19–11), (28–12), (25–13), of the vermis and an extension from it on either side called the **flocculus** (26–12), (24–13).

Figure 12

Superior and inferior views of the cerebellum.

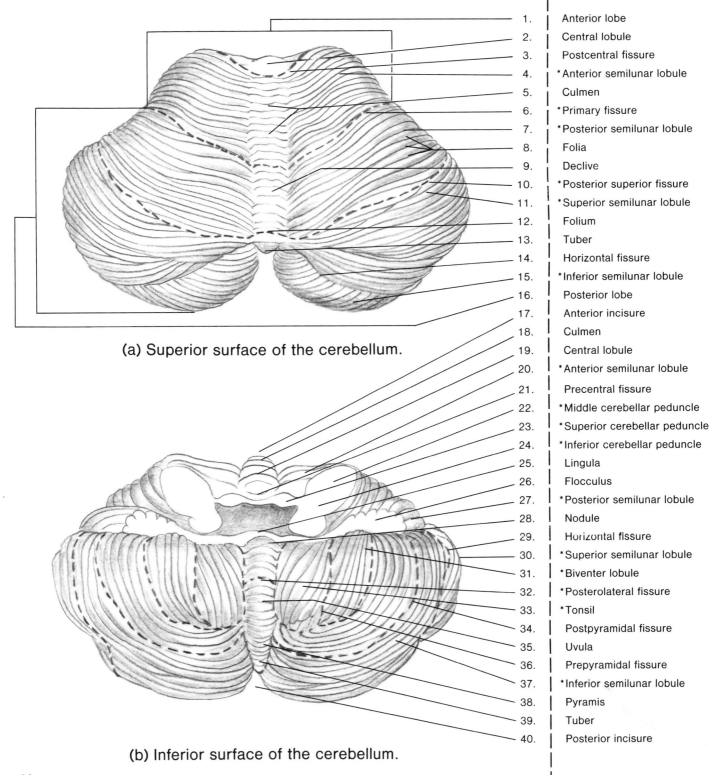

(a) Superior surface of the cerebellum.

(b) Inferior surface of the cerebellum.

1.   Anterior lobe
2.   Central lobule
3.   Postcentral fissure
4.   *Anterior semilunar lobule
5.   Culmen
6.   *Primary fissure
7.   *Posterior semilunar lobule
8.   Folia
9.   Declive
10. *Posterior superior fissure
11. *Superior semilunar lobule
12. Folium
13. Tuber
14. Horizontal fissure
15. *Inferior semilunar lobule
16. Posterior lobe
17. Anterior incisure
18. Culmen
19. Central lobule
20. *Anterior semilunar lobule
21. Precentral fissure
22. *Middle cerebellar peduncle
23. *Superior cerebellar peduncle
24. *Inferior cerebellar peduncle
25. Lingula
26. Flocculus
27. *Posterior semilunar lobule
28. Nodule
29. Horizontal fissure
30. *Superior semilunar lobule
31. *Biventer lobule
32. *Posterolateral fissure
33. *Tonsil
34. Postpyramidal fissure
35. Uvula
36. Prepyramidal fissure
37. *Inferior semilunar lobule
38. Pyramis
39. Tuber
40. Posterior incisure

Figure 13

Schematic of the lobes and fissures of the cerebellum.

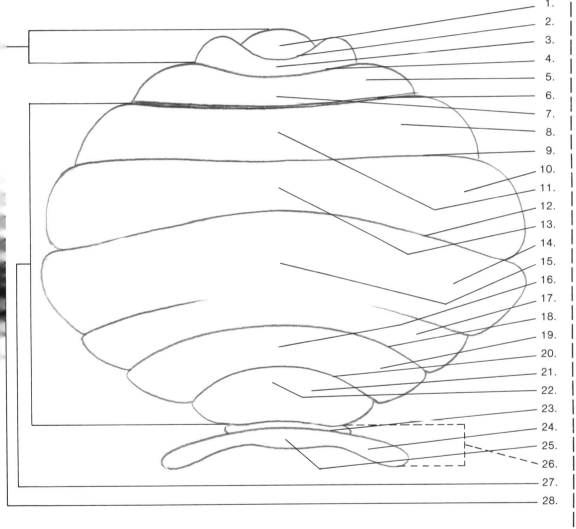

1. Lingula
2. Precentral fissure
3. Central lobule
4. Postcentral fissure
5. *Anterior semilunar lobule
6. *Primary fissure
7. Culmen
8. *Posterior semilunar lobule
9. *Posterior superior fissure
10. *Superior semilunar lobule
11. Declive
12. Horizontal fissure
13. Folium
14. *Inferior semilunar lobule
15. Tuber
16. Pyramis
17. Gracile lobule
18. Postpyramidal fissure
19. *Biventer lobule
20. Prepyramidal fissure
21. *Tonsil
22. Uvula
23. *Posterolateral fissure
24. Flocculus
25. Nodule
26. Flocculonodular lobe
27. Posterior lobe
28. Anterior lobe

THE SPINAL CORD: EXTERNAL FEATURES

The **spinal cord,** or *medulla spinalis,* occupies the upper two-thirds of the vertebral canal (see Fig. 14). Rostrally it is continuous with the medulla oblongata. Caudally it ends in the tapered **conus medullaris** (10–14). In the adult male it has a total length of about 45 centimeters (18 inches) and extends from the upper border of the atlas (the first cervical vertebra on which the skull rests) to the lower border of the first lumbar vertebra. A thin strip of connective tissue, the **filum terminale** (11–14), runs from the tip of the conus medullaris to the base of the **coccyx** (7–14) and attaches to the coccygeal ligament.

Until the third month of fetal life the spinal cord fills the entire vertebral canal. After that time, however, the vertebral column lengthens more rapidly than does the cord. At birth the cord ends opposite the third lumbar vertebra; it gradually recedes to the adult position during childhood.

The spinal cord is not perfectly cylindrical in shape. It is somewhat flattened dorsoventrally, particularly in the cervical region.[1] It has two prominent enlargements in those areas from which the nerves to the extremities exit. The **cervical enlargement** (2–14) includes the area extending from the third cervical to the second thoracic vertebra. The **lumbar enlargement** (9–14), which is somewhat less pronounced, begins at about the level of the ninth thoracic vertebra, reaches its largest diameter at the level of the twelfth thoracic vertebra, and then tapers rapidly into the **conus medullaris** (1,10–14). Viewed from the side (see Fig. 14c) the spinal cord is not straight. It has a forward concave curvature in the cervical region, a convex curvature backward in the thoracic region, and a concave curvature again in the lumbar section.

The surface of the spinal cord has a number of longitudinal furrows that can best be seen in the cross section of the cord (see Fig. 16). The **ventral median fissure** (3–14) is shown in Figure 14, as are the **ventrolateral sulci** (8–14).

Thirty-one pairs of spinal nerves emerge from the spinal cord and exit through the **intervertebral foramina** (14–14) of the spinal column. The spinal nerves are named to correspond to the region of the column from which they emerge. There are eight **cervical nerves** (5–14), twelve **thoracic nerves** (6–14), five **lumbar nerves** (12–14), five **sacral nerves** (15–14), and one **coccygeal nerve** (16–14).

Because the spinal canal is significantly longer than the spinal cord, it is obvious that the segments of the cord must lie higher than the corrsponding segments of the spinal column. The length of the nerve roots between the point of exit from the cord and the point of exit from the vertebrae increases from the more rostral to caudal. In the lumbar and sacral regions the spinal nerves travel together below the extent of the cord before they reach their foramina. This collection of nerve roots is referred to as the **cauda equina** (13–14), or *horse's tail.* In a spinal tap, or spinal puncture, the needle is always introduced into the subarachnoid space below the termination of the spinal cord itself.

---

[1] In Figures 14, 15, and 16, it is particularly important to remember that the terms *ventral* and *anterior* are equivalent, as are *dorsal* and *posterior.* We have generally used *dorsal* and *ventral* in this section, but many texts use *anterior* and *posterior.*

Figure 14

External aspects of the spinal cord.

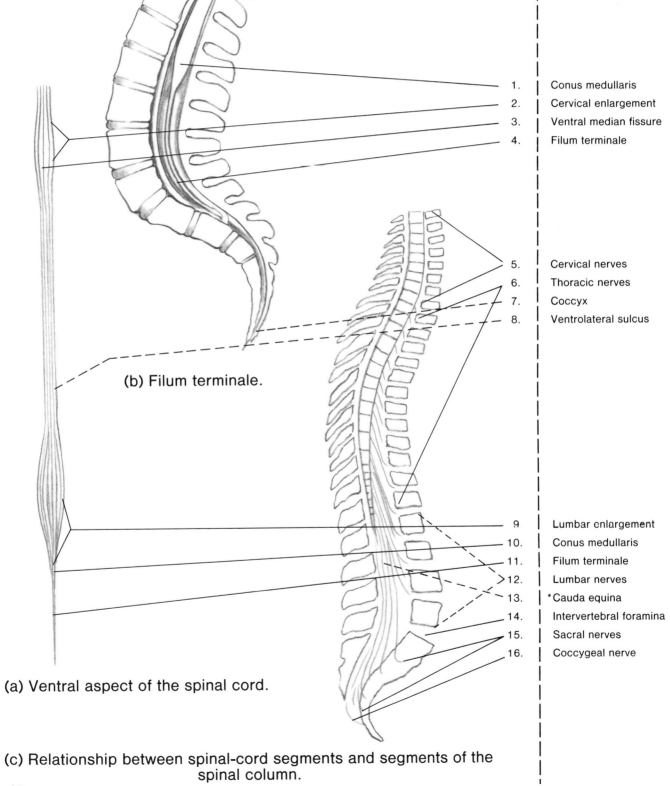

1. Conus medullaris
2. Cervical enlargement
3. Ventral median fissure
4. Filum terminale

5. Cervical nerves
6. Thoracic nerves
7. Coccyx
8. Ventrolateral sulcus

9 Lumbar enlargement
10. Conus medullaris
11. Filum terminale
12. Lumbar nerves
13. *Cauda equina
14. Intervertebral foramina
15. Sacral nerves
16. Coccygeal nerve

(b) Filum terminale.

(a) Ventral aspect of the spinal cord.

(c) Relationship between spinal-cord segments and segments of the spinal column.

SPINAL NERVE ROOTS

As illustrated in Figure 15, each of the 31 spinal nerves consists of a **dorsal root** (10,17–15) and a **ventral root** (6, 25–15), which join as they pass through the **intervertebral foramina** (14–14). The ventral nerve root is composed of efferent, or motor, fibers that arise from cell bodies located in the anterior and lateral columns of the gray matter of the spinal cord (see Fig. 15b). These fibers become medullated a short distance from their origins. They emerge from the cord in two or three irregular rows called **fila** (5–15) over an area that is about 3 millimeters wide.

The dorsal nerve roots come from cell bodies located in a **spinal ganglion** (11,18–15). They enter the cord at the dorsolateral sulcus through six or eight **fila.** Most of the spinal ganglia are located within the intervertebral foramina. The spinal ganglia of the sacral and coccygeal nerves, however, lie within the vertebral canal.

The drawing of the cross section of the spinal cord and nerve roots (Fig. 15b) shows some of the supporting structures of the cord as well as meninges closely attached to it. The meninges will be dicussed in more detail later (see Sect. 42). Here we are concerned only with their relationship to the spinal cord. The **dura mater** (15–15) consists of a tough fibrous tissue, which is a sheath enclosing the cord from the atlas to the level of the second sacral vertebra, where it covers the **filum terminale** (11–14). The spinal dura also encloses the dorsal and ventral roots of the spinal nerves.

The **arachnoid membrane** (13–15), which is continuous with the cranial arachnoid, lies inside the dura and follows it in all of its ramifications. The **pia mater** (16–15) invests the cord and nerve roots so closely that it cannot be pulled free. It is a very thin membrane of connective tissue. The sizeable area between the pia mater and the arachnoid is the **subarachnoid space** (23–15). This space contains the **cerebrospinal fluid.** The **subarachnoid septa** (14–15) are composed of projections, or *trabeculae,* from the pia mater and the arachnoid. The **denticulate ligaments** (19–15) are derived from the pia mater. They pass through the arachnoid and are attached to the dura. These ligaments are of structural significance and prevent the cord from being displaced.

Beyond the foramina the spinal nerve divides into several branches. The **dorsal ramus** (20–15) innervates the skin and muscles of the back, while the **ventral ramus** (22–15) innervates the ventral portion of the body. The **communicating rami** (24–15) branch from the ventral ramus and go to the sympathetic ganglia.

Figure 15

Spinal nerve roots.

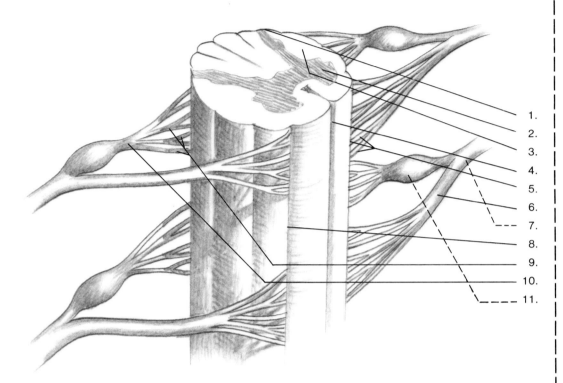

1.	Dorsolateral sulcus
2.	Gray matter
3.	White matter
4.	Ventral median fissure
5.	Fila of ventral root
6.	Ventral root
7.	Spinal nerve
8.	Ventrolateral sulcus
9.	Fila of dorsal root
10.	Dorsal root
11.	Spinal ganglion

(a) Ventral view of the spinal cord showing the nerve roots and fila.

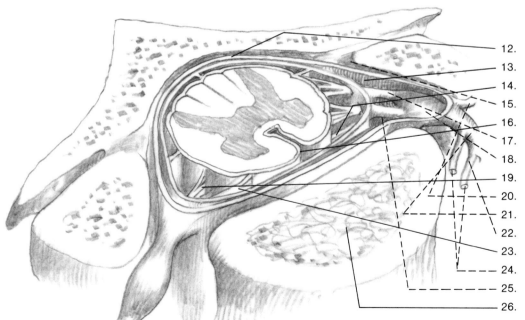

12.	Epidural space
13.	Arachnoid membrane
14.	Subarachnoid septa
15.	Dura mater
16.	Pla mater
17.	Dorsal root
18.	Spinal ganglion
19.	Denticulate ligament
20.	Dorsal ramus
21.	Spinal nerve
22.	Ventral ramus
23.	Subarachnoid space
24.	Communicating rami
25.	Ventral root
26.	Body of vertebra

(b) Cross section of the spinal cord showing the nerve roots and meninges.

THE SPINAL CORD: TRANSVERSE SECTION

A transverse section of the cord reveals that there are distinct white and gray areas (Fig. 16). The gray portion is centrally located and is shaped like a butterfly or the letter "H." The **central canal** (19–16) is located in the center of the gray matter. The white matter, composed of medullated fibers, surrounds the gray and has a number of identifiable fissures and sulci. The cord is divided into right and left halves by the **ventral median fissure** (27–16) and the **dorsal median sulcus** (3–16). The first penetrates the cord about 3 millimeters. The second is continuous with a thin sheet of neuroglia called the **dorsal median septum** (4–16), which extends more than halfway into the cord. The **dorsolateral sulcus** (6–16) is a distinct furrow into which the dorsal nerve roots enter. The **ventrolateral sulcus** (24–16), however, is less distinct since the ventral roots emerge in groups of filaments. Another furrow, the **dorsal intermediate sulcus** (2–16), which lies between the dorsal median sulcus and the dorsolateral sulcus, is found only in the cervical and upper thoracic cord. There are also numerous points at which fibrous tissue makes shallow entry into the cord. These are **glial trabeculae** (26–16).

The gray matter of the cord is composed of cell bodies, their dendrites, unmylenated nerve fibers, and neuroglia. It can be readily divided into parts that have distinct structural and functional significance. The central canal divides the crossbar into the **ventral gray commissure** (20–16) and the **dorsal gray commissure** (21–16). These commissures contain fibers that connect the two halves of the gray substance. The dorsal portion of the gray substance is known as the **dorsal horn** (12–16) or, frequently, the *dorsal column.* The dorsal horn can be further divided into an **apex** (10–16), a **caput** (11–16), or *head,* a **cervix** (13–16), and a **base** (14–16). The dorsal horn extends nearly to the surface of the cord and is separated from it by an area of white matter, the **dorsolateral fasciculus** (7–16), or *tract of Lissauer.* The ventral gray matter is called the **ventral horn** (23–16) or *ventral column.* In the thoracic level of the cord there is a third horn between the dorsal and ventral horns, the **lateral horn** (16–16) or *lateral column.* The head of the dorsal horn is capped with a material that stains lightly in certain preparations and is called the **substantia gelatinosa** (9–16). Just dorsolateral to it is a small crescent-shaped area called the **zona spongiosa** (8–16). Between the base of the dorsal horn and the ventral horn lies the **intermediate gray.** In the area just dorsal to the lateral horn is a lacy area of mixed white and gray substance called the **reticular substance** (15–16).

The white matter of the cord can be conveniently divided into three columns or funiculi. The **dorsal funiculus** (5–16) lies between the dorsal median septum and the dorsal roots. The **lateral funiculus** (25–16) is located between the dorsal and ventral roots, and the **ventral funiculus** (28–16) is between the ventral horn and the ventral median fissure. There is no clear line of demarcation between the ventral and lateral funiculi as there is between the dorsal and lateral. The funiculi of the two halves of the cord are connected dorsally by the **dorsal white commissure** (18–16) and ventrally by the **ventral white commissure** (22–16).

As can be readily seen in Figure 16 the relationship between the white and gray matter in the cord varies significantly from a transverse section at one level to a transverse section at another level. In general, as one proceeds from a caudal to a rostral location, there is a larger proportion of white to gray. The cervical portion contains the largest number of fibers, thus the greatest proportion of white matter. In the area of the cervical and lumbar enlargements the two horns of gray matter are significantly larger because of the larger nerves of the extremities. In the thoracic segment of the cord the nerve cells of the preganglionic fibers of the sympathetic system expand the size of the lateral horn.

Figure 16

Transverse sections through the spinal cord.

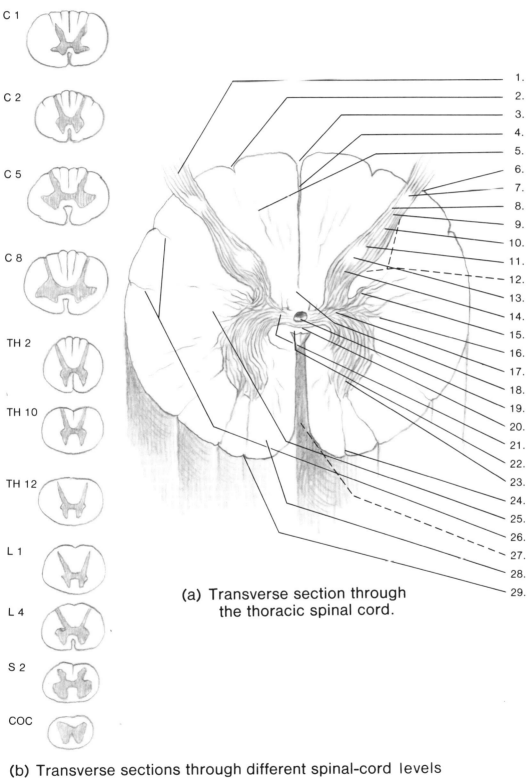

C 1

C 2

C 5

C 8

TH 2

TH 10

TH 12

L 1

L 4

S 2

COC

1.	Dorsal nerve root
2.	Dorsal intermediate sulcus
3.	Dorsal median sulcus
4.	Dorsal median septum
5.	Dorsal funiculus
6.	Dorsolateral sulcus
7.	*Dorsolateral fasciculus
8.	Zona spongiosa
9.	Substantia gelatinosa
10.	Apex
11.	*Caput
12.	*Dorsal horn
13.	Cervix
14.	Base
15.	*Reticular substance
16.	*Lateral horn
17.	Intermediate horn
18.	Dorsal white commissure
19.	Central canal
20.	Ventral gray commissure
21.	Dorsal gray commissure
22.	Ventral white commissure
23.	*Ventral horn
24.	Ventrolateral sulcus
25.	Lateral funiculus
26.	Glial trabeculae
27.	Ventral median fissure
28.	Ventral funiculus
29.	Ventral nerve roots

(a) Transverse section through
the thoracic spinal cord.

(b) Transverse sections through different spinal-cord levels
showing variations in white and gray matter.

DERMATOMES

The dorsal roots of the spinal cord contain the fibers of sensory nerves that carry the impulses for touch, pain, and temperature. Each spinal nerve receives impulses from a particular area of the skin. Those skin areas supplied by a single nerve root are called **dermatomes.** Any given skin area, except the back of the head, is innervated by more than one spinal nerve—usually three and sometimes four. Thus, damage to one dorsal nerve root does not result in complete anesthesia in any portion of the body. The overlapping of adjacent dermatomes is not the same for all of the cutaneous sensations. Evidence indicates that the overlap is greater for touch than it is for pain and temperature.

The dermatome map in Figure 17 is a composite of several others and will provide a general schematic that will help you visualize the sensory innervation patterns. The dermatome patterns have been developed on the basis of several different kinds of studies, including the following: study of sensory loss from herniation of intervertebral disks, measurements of skin resistance, study of anesthesia resulting from resections of dorsal roots to relieve spastic conditions, and study of hyperalgesia associated with Herpes zoster, a disease which frequently attacks a single spinal ganglion.

As the diagram shows, the relationship between the spinal nerves and the dermatomes of the trunk area is neat and orderly. Each dermatome is a band encircling the body from the middorsal to the midventral line, and each is associated with the spinal nerve in that immediate area. In the head area and in the extremities the situation is somewhat more complicated. This is due to the fact that the basic innervation of metameres, or body segments, occurs during early embryonic life before the limb buds have developed. During development the metameres migrate into the limb buds and are arranged parallel to the long axis of the future limb.

The segmental pattern for muscle innervation is quite similar to the dermatome pattern. As is true for the cutaneous innervation, most of the muscles of the body are innervated by two, three, or even four ventral roots. Thus, damage to a single ventral nerve root does not result in loss of muscle function, but rather in a weakness in the affected area.

The dermatome pattern is of obvious clinical significance in localizing the level and particular site of spinal pathology.

Figure 17

Pattern of cutaneous innervation
of the dermatomes.

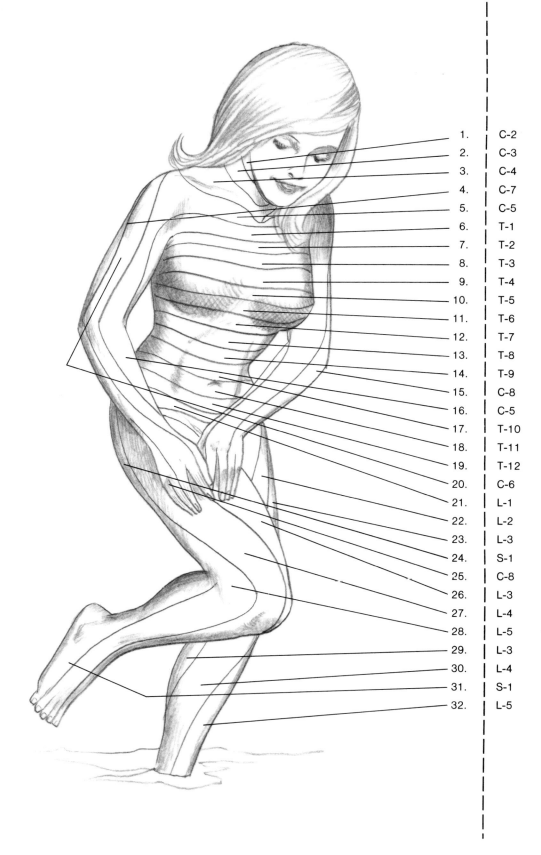

1.	C-2
2.	C-3
3.	C-4
4.	C-7
5.	C-5
6.	T-1
7.	T-2
8.	T-3
9.	T-4
10.	T-5
11.	T-6
12.	T-7
13.	T-8
14.	T-9
15.	C-8
16.	C-5
17.	T-10
18.	T-11
19.	T-12
20.	C-6
21.	L-1
22.	L-2
23.	L-3
24.	S-1
25.	C-8
26.	L-3
27.	L-4
28.	L-5
29.	L-3
30.	L-4
31.	S-1
32.	L-5

SPINAL-CORD PATHWAYS

The nerve fibers that make up the white matter of the spinal cord are not distributed in a random manner. In general, the axons of nerves that have a similar function and place of origin and termination are collected together to form a reasonably discrete bundle. These bundles are known as *fiber tracts,* or *fasciculi.* While this is generally true, it should also be recognized that there is a significant intermingling of fibers in adjacent tracts. Thus, Figure 18 is a schematic, and the firm lines are somewhat of a misrepresentation. However, once that is recognized, the diagram is most useful for learning the relationships among the fiber systems.

The pathways of the fibers of more important fasciculi will be covered in greater detail in a later section dealing with the major afferent and efferent systems. In this section only brief comments will be made regarding origin, termination, and function.

In general, the fiber tracts form the conduction link between the brain and the effector mechanisms and between the body receptors and the higher centers of the brain. Some fiber tracts, however, the **association fibers,** serve to connect the different segments of the cord.

Fortunately for the student most of the names of the fasciculi are descriptive. The point of origin is given first and the point of termination second. Thus, most ascending tracts will have the prefix *spino-,* while descending tracts will have as a prefix some portion of the brain, for example, *cerebro-.* In addition, many tract names are preceded by a positional name that designates the tract's relative position in the cord. Thus, the lateral spinothalamic tract is located in the lateral funiculus, while the ventral spinothalamic is in the ventral funiculus. The dorsal and ventral spinocerebellar tracts are both in the lateral funiculus, but the dorsal spinocerebellar is posterior to the ventral. There are exceptions, but these rules are helpful heuristic devices for learning the spinal fasciculi.

Ascending Fasciculi

The **fasciculus gracilis** (4–18), or *tract of Goll,* carries muscle and joint sensations. It increases in size as it ascends and picks up more fibers from the upper segments of the body. It terminates in the nucleus gracilis in the medulla oblongata.

The **fasciculus cuneatus** (2–18), or *tract of Burdach,* also carries muscle and joint sensations and increases in size as it ascends. Like the fasciculus gracilis, it is derived from the posterior nerve roots of the thoracic and cervical regions. It terminates in the medulla in the cuneate and accessory cuneate nuclei.

The **dorsal spinocerebellar fasciculus** (3–18) conducts impulses from the leg and trunk muscles. It runs from the spinal cord through the inferior cerebellar peduncle and terminates in the cerebellum.

The specific functions of the **ventral spinocerebellar fasciculus** (10–18) are not known. It originates at the level of the third lumbar spinal nerves and terminates in the cerebellum by way of the medulla, pons, and anterior medullary velum.

The **lateral spinothalamic fasciculus** (13–18) transmits sensations of pain and temperature. It originates in the cells of the dorsal horn of the spinal-cord gray matter. The fibers cross to the opposite side of the cord through the anterior commissure and ascend to the ventral posterior lateral nucleus of the thalamus.

The **spinotectal fasciculus** (22–18) is probably concerned with the visual reflexes. It arises from the cells in the dorsal horn, crosses to the opposite side, and ascends to the superior colliculi of the midbrain.

The exact function of the **spinoolivary fasciculus** (24–18) is not known. The fibers run from the spinal cord to the inferior olivary nucleus of the medulla.

The **ventral spinothalamic fasciculus** (23–18) transmits sensations of touch. It originates in the cells of the posterior horn, and the fibers cross to the opposite side in the anterior commissure. The fibers end in the thalamus.

Descending Fasciculi

The **lateral cerebrospinal fasciculus** (9–18), or *crossed pyramidal tract,* is concerned with voluntary movement. The fibers originate in the large pyramidal cells of the precentral gyrus of the cerebral hemispheres and extend through the entire spinal cord. They cross to the opposite side of the cord at the pyramidal decussation. Most of them terminate in the dorsal horn cells, although a few go directly to the ventral horn cells.

The **rubrospinal fasciculus** (11–18) transmits impulses relating to the cerebellar reflexes. It originates in the red nucleus of the midbrain on the opposite side. The fibers descend as far as the sacral region.

The **reticulospinal fasciculus** (18–18) is also probably concerned with cerebellar reflexes and is a part of the reticular activating system. It originates in the reticular

(Continued)

Figure 18

Principal fasciculi of the spinal cord.

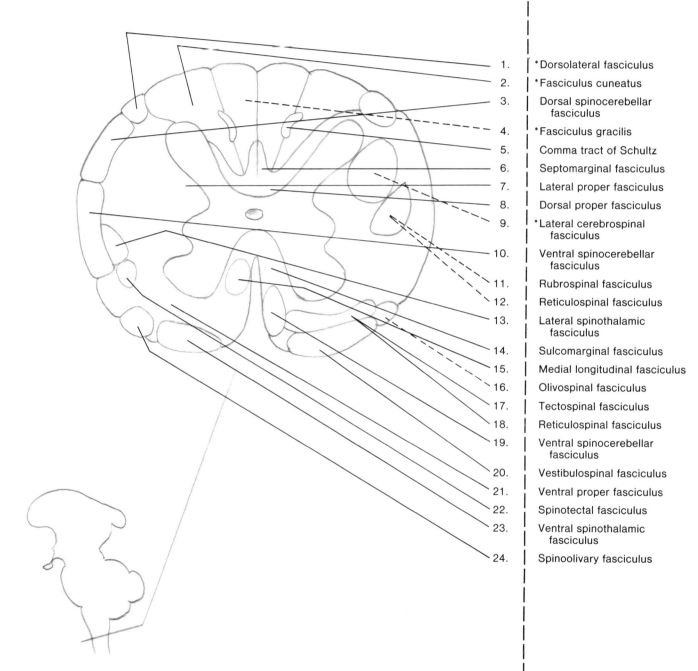

1.   *Dorsolateral fasciculus
2.   *Fasciculus cuneatus
3.   Dorsal spinocerebellar fasciculus
4.   *Fasciculus gracilis
5.   Comma tract of Schultz
6.   Septomarginal fasciculus
7.   Lateral proper fasciculus
8.   Dorsal proper fasciculus
9.   *Lateral cerebrospinal fasciculus
10.   Ventral spinocerebellar fasciculus
11.   Rubrospinal fasciculus
12.   Reticulospinal fasciculus
13.   Lateral spinothalamic fasciculus
14.   Sulcomarginal fasciculus
15.   Medial longitudinal fasciculus
16.   Olivospinal fasciculus
17.   Tectospinal fasciculus
18.   Reticulospinal fasciculus
19.   Ventral spinocerebellar fasciculus
20.   Vestibulospinal fasciculus
21.   Ventral proper fasciculus
22.   Spinotectal fasciculus
23.   Ventral spinothalamic fasciculus
24.   Spinoolivary fasciculus

substance of the tegmentum. Reticulospinal fibers travel with both the tectospinal fasciculus and the rubrospinal fasciculus.

The **tectospinal fasciculus** (17–18) is probably concerned with visual reflexes since the cells originate in the opposite superior colliculus of the midbrain. It terminates in the ventral horn cells.

The function of the **olivospinal fasciculus** (16–18) has not been established. It arises in cells of the inferior olive of the medulla. It is found only in the cervical region of the spinal cord.

The **vestibulospinal fasciculus** (20–18) transmits impulses related to equilibrium and antigravity reflexes. The fibers originate in the vestibular nuclei of the medulla and can be traced to the level of the sacral spinal nerves.

The **ventral cerebrospinal fasciculus**, or *direct pyramidal tract,* is concerned with voluntary movement. The fibers arise in the pyramidal cells of the motor area of the cortex and descend to the spinal cord on the same side just lateral to the sulcomarginal fasciculus, where they cross the cord through the anterior white commissure at the level of termination. This is a small tract not shown on the diagram.

The **medial longitudinal fasciculus** (15–18) is the principal connection between the vestibular complex and the centers in the medulla that coordinate head and eye movements. It descends to the second or third thoracic segments. It actually contains both ascending and descending fibers.

Association Fasciculi

The **proper fasciculi,** also called *ground bundles,* are bundles of fibers that connect the various levels of the cord. They are located next to the gray matter and are designated according to the funiculus in which they are found; thus, there are three sets of ground bundles: **dorsal proper fasciculus** (8–18), **lateral proper fasciculus** (7–18), and **ventral proper fasciculus** (21–18).

Other bundles of association fibers include the **comma tract of Schultz** (5–18), the **dorsolateral fasciculus** (1–18), or *tract of Lissauer,* the **septomarginal fasciculus** (6–18), and the **sulcomarginal fasciculus** (14–18).

THE INTERNAL MEDULLA: TRANSITION BETWEEN SPINAL CORD AND MEDULLA

The medulla oblongata is continuous with the spinal cord, and there is relatively little external change at the point of transition. The external changes more rostral to the transition have been illustrated in Figures 7, 8, and 9. A knowledge of the external medulla will help in understanding the internal changes.

In the internal medulla there is a progressively greater rearrangement of the fiber tracts and a progressively greater distribution of the gray matter from the more-caudal to the more-rostral areas. It can be readily seen that there is little resemblance in the transverse section of the medulla at the transition point between the cord and the medulla (Fig. 19) and at the transition point between the medulla and the pons (Fig. 24). The enlargement of the spinal canal into the fourth ventricle forces dorsal structures into a lateral position. A number of nuclei, including the olive, the nucleus gracilis, the nucleus cuneatus, the spinal-nerve nuclei, and other nuclei, tend to break up the fasciculi into patches of white and gray matter.

The next five drawings (Figs. 19–23) are transverse sections of the medulla, through which the changes can be traced. Detailed study of these illustrations will provide a thorough knowledge of the structures of the medulla and their interrelationships. No attempt is made at this point to trace the connections of the nuclei of the cranial nerves. These will be considered in detail in later sections (see Figs. 63, 64, and 65).

These drawings are idealized and do not represent actual histological preparations. All of the features presented could not be seen in any one type of preparation. These are teaching devices that will enable you to learn the relationships between the nuclei and the fasciculi. It should also be remembered that the boundaries of the fasciculi and the nuclei are never as neat and discrete as depicted in drawings. Also, there is considerable variability from one brain to another. Nervous systems differ just as the shapes of noses differ. By becoming familiar with the relative locations of the internal structures of the brain stem, you will be prepared to understand their relationships when function is discussed in later sections. In order to get a better idea of the shifting nature of the tracts and the nuclei, you may wish to trace particular ones through the different sections of the bain stem.

The gray matter of the medulla at the point just above the first cervical nerve (Fig. 19) is not greatly different from that in the cord. The same general shape is apparent, although there is more gray matter around the central canal than is found in the cord. The **dorsal horn** (10–19) is somewhat larger and begins to curve in a lateral direction around the **lateral cerebrospinal fasciculus** (18–19). The **substantia gelatinosa** (9–16) of the cord now becomes the **nucleus of the spinal tract of the trigeminal nerve [V]** (9–19). The **tract of the trigeminal nerve [V]** (2–19) is apparent just dorsolateral to the nucleus. At this level it is composed of both descending sensory fibers of the trigeminal nerve and ascending fibers of the **dorsolateral fasciculus** (1–18), or *tract of Lissauer.*

As in the spinal cord, the **posterior intermediate septum** (4–19) separates the **fasciculus cuneatus** (6–19) from the more medial **fasciculus gracilis** (5–19). In some preparations the **nucleus gracilis** (7–19) can be seen amid the fibers of the fasciculus gracilis at this level. The fasciculus cuneatus sends collateral fibers into the dorsal horn of the gray matter.

Just above the **ventral median fissure** (21–19) the beginning of the **pyramidal decussation** (16–19) can be seen. The **lateral cerebrospinal fasciculus** (18–19) has begun to divide into a large number of small fiber bundles separated by patches of gray matter. As a result the area has a generally mottled appearance. Some of the fibers from these bundles can be seen running in a ventromedial direction to begin the decussation of the pyramids. The fibers cross at the midline and continue in a rostral direction in the pyramidal tract of the opposite side.

As indicated in Figure 19, the other fasciculi have not been appreciably displaced from their relative positions in the spinal cord.

Figure 19

Transverse section through the medulla at the point of transition between the spinal cord and the medulla.

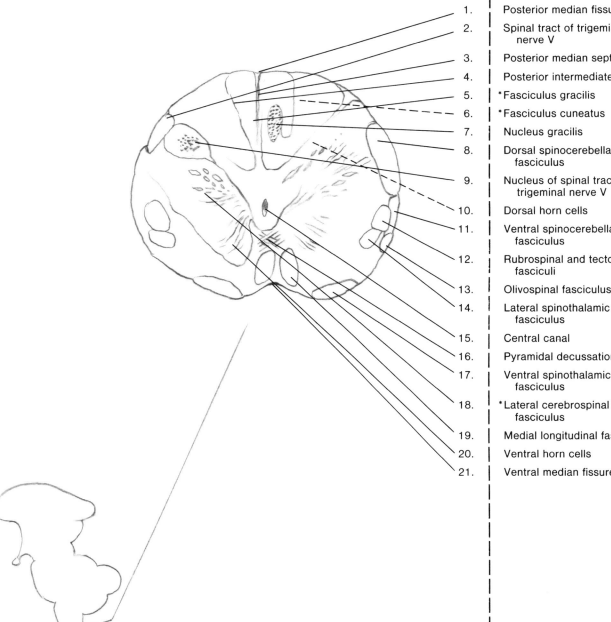

1. Posterior median fissure
2. Spinal tract of trigeminal nerve V
3. Posterior median septum
4. Posterior intermediate septum
5. *Fasciculus gracilis
6. *Fasciculus cuneatus
7. Nucleus gracilis
8. Dorsal spinocerebellar fasciculus
9. Nucleus of spinal tract of trigeminal nerve V
10. Dorsal horn cells
11. Ventral spinocerebellar fasciculus
12. Rubrospinal and tectospinal fasciculi
13. Olivospinal fasciculus
14. Lateral spinothalamic fasciculus
15. Central canal
16. Pyramidal decussation
17. Ventral spinothalamic fasciculus
18. *Lateral cerebrospinal fasciculus
19. Medial longitudinal fasciculus
20. Ventral horn cells
21. Ventral median fissure

THE INTERNAL MEDULLA: PYRAMIDAL DECUSSATION

The most obvious feature of a transverse section of the medulla at the level illustrated in Figure 20 is the band of fibers crossing just ventral to the **central canal** (11–20) to form the **pyramidal decussation** (15–20). The **pyramids** (22–20) on either side of the **ventral median fissure** (23–20) have become much more prominent. The pyramids and the pyramidal decussation have shifted the central canal in a dorsal direction, and the ventral gray area found in more-caudal sections has essentially disappeared and been replaced by the fibers of the decussation. However, there is an increase in the amount of gray matter called the **central gray** (8–20), which lies dorsal to the canal. The **lateral cerebrospinal fasciculus** (9–18), (18–19) has been essentially replaced by the **reticular substance** (9–20), which is the most caudal portion of the *brain-stem reticular formation.* The **dorsal spinocerebellar fasciculus** (10–20) and the **ventral spinocerebellar fasciculus** (12–20) have not shifted their relative positions from those held in the previous section. This is also true of the **spinal tract of the trigeminal nerve [V]** (3–20) and its **nucleus** (6–20).

Other changes are taking place in the dorsal portion of the medulla. The **nucleus gracilis** (5–20) is larger, and the **nucleus cuneatus** (7–20) has appeared. The **dorsal median sulcus** (1–20) is more prominent, and the separation of the **fasciculus gracilis** (4–20) and **fasciculus cuneatus** (2–20) is more definite because of the wider **dorsal intermediate septum,** which lies in the dorsal intermediate sulcus. In more-rostral sections, the nucleus gracilis and nucleus cuneatus become larger, and their respective fasciculi become smaller as increasingly more of the fibers terminate in the nuclei.

The increase in size of the pyramids has pushed the **vestibulospinal fasciculus** (19–20) in a more lateral direction. Several other fasciculi have had their courses only moderately shifted, as Figure 20 shows.

The **nucleus of the spinal root of the accessory nerve [XI]** (14–20) becomes apparent in this section somewhat lateral to the midline. The **supraspinal nucleus** (18–20), which receives fibers from the **hypoglossal nucleus,** is somewhat dorsal and lateral to the ventral medial septum in the ventral median fissure.

Figure 20

Transverse section through the medulla at the point of the pyramidal decussation.

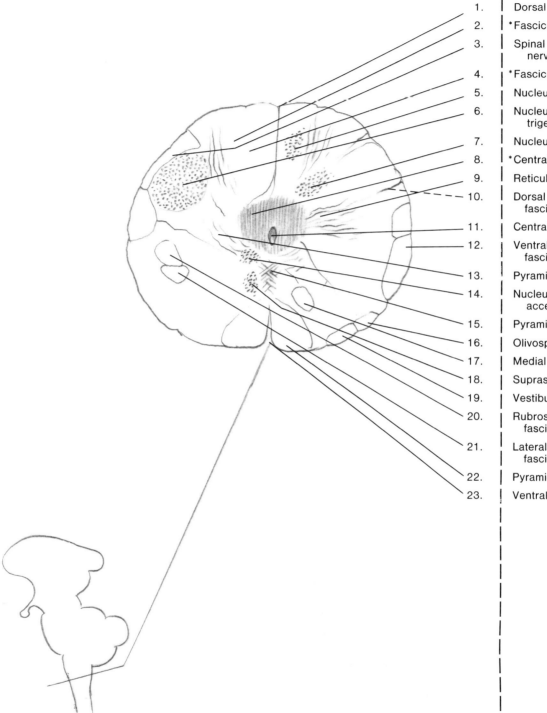

1. Dorsal median sulcus
2. *Fasciculus cuneatus
3. Spinal tract of trigeminal nerve V
4. *Fasciculus gracilis
5. Nucleus gracilis
6. Nucleus of spinal tract of trigeminal nerve V
7. Nucleus cuneatus
8. *Central gray
9. Reticular substance
10. Dorsal spinocerebellar fasciculus
11. Central canal
12. Ventral spinocerebellar fasciculus
13. Pyramidal fibers
14. Nucleus of spinal root of accessory nerve XI
15. Pyramidal decussation
16. Olivospinal fasciculus
17. Medial longitudinal fasciculus
18. Supraspinal nucleus
19. Vestibulospinal fasciculus
20. Rubrospinal and tectospinal fasciculi
21. Lateral spinothalamic fasciculus
22. Pyramid
23. Ventral median fissure

THE INTERNAL MEDULLA: MEDIAL LEMNISCUS DECUSSATION

Several new features are introduced at the level of the medulla illustrated in Figure 21. The most obvious of these is the **decussation of the medial lemniscus (22–21)**, or *sensory decussation*. The fibers forming the decussation originate in the **nucleus gracilis (5–21)** and the **nucleus cuneatus (7–21)** of the opposite side. They cross at the midline and turn in a rostral direction, forming the fiber tract called the **medial lemniscus (24–21)**, which ultimately terminates in the **ventral posterior lateral nucleus of the thalamus (33–38)**. The bands of crossing fibers are the **internal arcuate fibers (10–21)**. The decussation of the medial lemniscus determines that the sensory representation for each half of the body will appear on the contralateral side of the cerebral cortex. Thus, damage to the medial lemniscus results in dysfunctions of the tactile and kinesthetic senses on the opposite side of the body. The **medial longitudinal fasciculus (19–21)** and the **tectospinal tract (20–21)** have been displaced dorsally and lie between the medial lemniscus and the **central canal (13–21)**.

At this point in the medulla most of the fibers of the **fasciculus gracilis (3–21)** and the **fasciculus cuneatus (1–21)** have terminated in their respective nuclei, although the fasciculus cuneatus is larger than the gracilis. Lateral to the cuneate nucleus is the **accessory cuneate nucleus (9–21)**. Fibers arising from it go through the **dorsal external arcuate fibers (12–21)** and the inferior cerebellar peduncle to the cerebellum.

Within the **ventral median fissure (30–21)** can be seen the **ventral external arcuate fibers (31–21)** as they extend from that point across the ventral medulla to the **arcuate nuclei (28–21)**, which are on the ventral aspect of the **pyramids (27–21)**.

The most caudal portion of the **medial accessory olivary nucleus (23–21)** can be seen just lateral to the medial lemniscus. The lowest tip of the **hypoglossal nucleus** is located in the ventral portion of the **central gray (6–21)**. The **dorsal nucleus of the vagus nerve [X] (11–21)** lies in the lateral central gray, and the **commissural nucleus of the vagus nerve [X] (8–21)** is dorsomedial to it. The fibers of the **hypoglossal nerve [XII] (26–21)** run from the nucleus along the lateral portion of the medial lemniscus. The **spinal tract of the trigeminal nerve [V] (2–21)** and the **nucleus of the spinal tract of the trigeminal nerve [V] (4–21)** remain in the same relative positions as in Section 18.

The **reticular substance (21–21)** at this level contains fairly definite groups of cells and is traversed by many of the internal arcuate fibers. One of the reticular nuclei, the **lateral reticular nucleus (29–21)**, is located in the ventrolateral portion of the reticular substance. The **ventral reticular nucleus (17–21)** can also be seen in this section.

The **nucleus ambiguus (25–21)** is apparent in this section, and the **supraspinal nucleus (18–21)** is in the same relative position as at the level of the pyramidal decussation.

Figure 21

Transverse section through the medulla at the point of the decussation of the medial lemniscus.

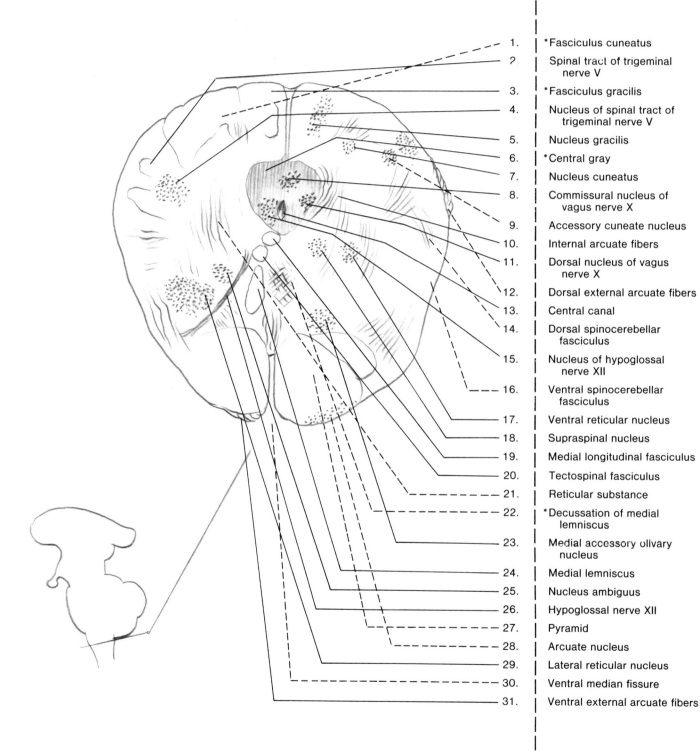

1.   *Fasciculus cuneatus
2.   Spinal tract of trigeminal nerve V
3.   *Fasciculus gracilis
4.   Nucleus of spinal tract of trigeminal nerve V
5.   Nucleus gracilis
6.   *Central gray
7.   Nucleus cuneatus
8.   Commissural nucleus of vagus nerve X
9.   Accessory cuneate nucleus
10. Internal arcuate fibers
11. Dorsal nucleus of vagus nerve X
12. Dorsal external arcuate fibers
13. Central canal
14. Dorsal spinocerebellar fasciculus
15. Nucleus of hypoglossal nerve XII
16. Ventral spinocerebellar fasciculus
17. Ventral reticular nucleus
18. Supraspinal nucleus
19. Medial longitudinal fasciculus
20. Tectospinal fasciculus
21. Reticular substance
22. *Decussation of medial lemniscus
23. Medial accessory olivary nucleus
24. Medial lemniscus
25. Nucleus ambiguus
26. Hypoglossal nerve XII
27. Pyramid
28. Arcuate nucleus
29. Lateral reticular nucleus
30. Ventral median fissure
31. Ventral external arcuate fibers

THE INTERNAL MEDULLA: INFERIOR OLIVE

The distinctive outline of the **inferior olivary nucleus** (24–22) can be seen at the level illustrated in Figure 22. The convoluted gray matter surrounds bands of white fibers except for an opening in the medial portion called the **hilus** (30–22). The band of fibers emerging from the olivary nucleus in a medial direction is called the **olivary peduncle** (33–22). These fibers cross the midline and proceed through the opposite olivary nucleus or around it, going in a dorsal direction through the **spinal tract of the trigeminal nerve [V]** (13–22) and into the **inferior cerebellar peduncle** (10–22), or *restiform body*.

There are two masses of gray matter associated with the inferior olivary nucleus: the **medial accessory olivary nucleus** (32–22), which lies between the hilus of the inferior olive and the medial lemniscus, and the **dorsal accessory olivary nucleus** (22–22), which lies just posterior to the inferior olive. The principal nucleus of the inferior olive is surrounded by a large bundle of mylenated fibers that come from the cerebral cortex, the red nucleus, and the periaqueductal gray of the midbrain and that terminate on the cells of the olive. These fibers are referred to collectively as the **amiculum olivae** (38–22).

The expansion of the olivary nucleus at this point in the medulla has forced a number of the fasciculi, including the **ventral spinocerebellar fasciculus** (16–22), the **spinothalamic fasciculus** (20–22), and the **rubrospinal fasciculus** (17–22), into a more dorsal position.

The **medial longitudinal fasciculus** (18–22) lies at the midline, ventral to the **nucleus of the hypoglossal nerve [XII]** (14–22). The fibers of the **hypoglossal nerve [XII]** (29–22) run to the **medial lemniscus**. The **tectospinal tract** (27–22) is a bundle of fibers descending from the superior colliculi and located in the ventral portion of the medial longitudinal fasciculus.

The central canal has expanded into the **fourth ventricle** (2–22). The **nucleus gracilis** (4–22) has almost disappeared and its remaining outlines are indefinite. The **inferior cerebellar peduncle** (10–22) has increased in size and lies lateral to the **lateral cuneate nucleus** (7–22) and the **spinal tract of the trigeminal nerve [V]** (13–22). The **nucleus of the spinal tract of the trigeminal nerve [V]** (11–22) is somewhat broken by the fibers running from the olivary nucleus to the inferior cerebellar peduncle.

The **nucleus intercalatus** (6–22) lies dorsomedial to the hypoglossal nucleus, and lateral to that is the **dorsal motor nucleus of the vagus nerve [X]** (9–22). The **tractus solitarius** (5–22) and its nucleus are lateral to the motor nucleus of the vagus.

The **pyramid** (35–22), the **arcuate nucleus** (36–22), and the **ventral external arcuate fibers** (37–22) occupy the same relative positions they did at the level of the medial lemniscus decussation.

The **lateral reticular nucleus** (21–22) is just medial to the ventral spinocerebellar fasciculus, and the **internal arcuate fibers** (26–22) can be seen coursing through the reticular substance. Dorsolateral to the lateral reticular nucleus is the **nucleus ambiguus** (19–22), in which motor fibers originate. They run through the **accessory [XI]**, **glossopharyngeal [IX]**, and **vagus [X]** nerves.

Two other nuclear masses of the reticular formation appear at this level of the medulla. The **nucleus reticularis gigantocellularis** (28–22) is located lateral to the medial longitudinal fasciculus; fibers descend from this nucleus to form the reticulospinal fasciculus. The **paramedian reticular nuclei** (23–22) are ventral and medial to the **gigantocellularis nuclei**; fibers project from this nucleus to the cerebellum. The **nucleus of Roller** (15–22) is a collection of cells just ventral to the hypoglossal nucleus that sends fibers into the reticular formation.

The **medial vestibular nucleus** (3–22) and the **inferior vestibular nucleus** (8–22) are visible in the dorsal aspect of this section. The **pontobulbar nucleus** (1–22) in the dorsolateral area is continuous with the nuclei of the pons, the **pontine nuclei** (23–24).

Near the midline of this section lies a group of decussating fibers called the **median raphe** (34–22). The cells mingled with these fibers constitute the **nucleus of the raphe** (25–22).

Figure 22

Transverse section of the medulla at the level of the middle of the inferior olivary nucleus.

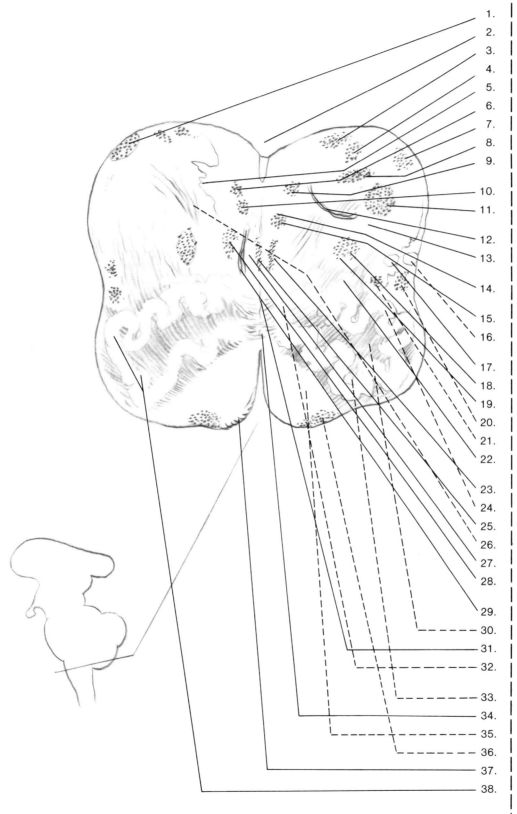

1. Pontobulbar nucleus
2. Fourth ventricle
3. *Medial vestibular nucleus
4. Nucleus gracilis
5. Tractus solitarius and nucleus
6. Nucleus intercalatus
7. Lateral cuneate nucleus
8. Inferior vestibular nucleus
9. Dorsal motor nucleus of vagus nerve X
10. *Inferior cerebellar peduncle
11. Nucleus of spinal tract of trigeminal nerve V
12. Vagus nerve X
13. Spinal tract of trigeminal nerve V
14. Nucleus of hypoglossal nerve XII
15. Nucleus of Roller
16. Ventral spinocerebellar fasciculus
17. Rubrospinal fasciculus
18. Medial longitudinal fasciculus
19. Nucleus ambiguus
20. Spinothalamic fasciculus
21. Lateral reticular nucleus
22. Dorsal accessory olivary nucleus
23. Paramedian reticular nucleus
24. Inferior olivary nucleus
25. Nucleus of raphe
26. Internal arcuate fibers
27. Tectospinal fasciculus
28. Nucleus reticularis gigantocellularis
29. Hypoglossal nerve XII
30. Hilus of olivary nucleus
31. Medial lemniscus
32. Medial accessory olivary nucleus
33. Olivary peduncle
34. Median raphe
35. Pyramid
36. Arcuate nucleus
37. Ventral external arcuate fibers
38. Amiculum olivae

THE INTERNAL MEDULLA: COCHLEAR NUCLEI

The **inferior cerebellar peduncle** (1–23), or *restiform body,* is prominent at the level of the internal medulla illustrated in Figure 23, and the **olivocerebellar fibers** (13–23) can be seen entering it from the **inferior olivary nucleus** (20–23). The **dorsal cochlear nucleus** (8–23) arches over the peduncle; just medial to it is the **inferior vestibular nucleus** (7–23), and more medial to it, the **medial vestibular nucleus** (6–23).

At this level the **hypoglossal nucleus** and the **nucleus intercalatus** have disappeared and the **nucleus of the eminentia teres** (4–23) is in that relative position. Just ventral to it is the **nucleus prepositus hypoglossi** (5–23). Fibers of the **stria medullaris** (2–23) can be seen at the edge of the **fourth ventricle** (3–23). The **medial longitudinal fasciculus** (10–23) and the **tectospinal tract** (17–23) are in the same relative positions as at the level of the inferior olive.

The **inferior olivary nucleus** (20–23) is still a prominent feature of the medulla, and the **hilus** (23–23) is clearly discernible. The **pyramid** (24–23) is ventral to the olive and has **ventral external arcuate fibers** (27–23) coursing around its medial and ventral aspects and emerging from the **ventral median fissure** (28–23). The **arcuate nucleus** (25–23) is located at the edge of the ventral portion of the pyramid.

The **vagus nerve [X]** (14–23) emerges at a point just below the **spinal tract** and the **nucleus of the spinal tract of the trigeminal nerve [V]** (11–23). The **nucleus ambiguus** (19–23) is located medial to the point of emergence of the vagus nerve.

The **lateral reticular nucleus** (15–23) is just medial to the **anterior spinocerebellar tract** (26–23). The **medial lemniscus** (22–23) lies next to the **median raphe** (18–23) just above the hilus of the olive. The **nucleus reticularis gigantocellularis** (12–23), the **tractus solitarius** (9–23) and its **nucleus,** and the **nucleus of the raphe** (16–23) are in the same relative positions in this section as at the level of the inferior olive.

Figure 23

Transverse section of the medulla at the level of the cochlear nuclei.

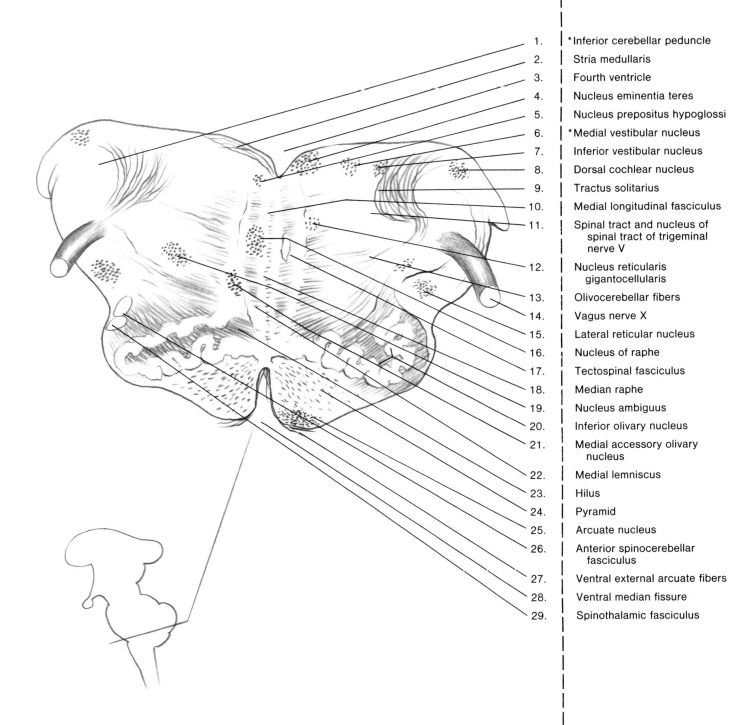

1. *Inferior cerebellar peduncle
2. Stria medullaris
3. Fourth ventricle
4. Nucleus eminentia teres
5. Nucleus prepositus hypoglossi
6. *Medial vestibular nucleus
7. Inferior vestibular nucleus
8. Dorsal cochlear nucleus
9. Tractus solitarius
10. Medial longitudinal fasciculus
11. Spinal tract and nucleus of spinal tract of trigeminal nerve V
12. Nucleus reticularis gigantocellularis
13. Olivocerebellar fibers
14. Vagus nerve X
15. Lateral reticular nucleus
16. Nucleus of raphe
17. Tectospinal fasciculus
18. Median raphe
19. Nucleus ambiguus
20. Inferior olivary nucleus
21. Medial accessory olivary nucleus
22. Medial lemniscus
23. Hilus
24. Pyramid
25. Arcuate nucleus
26. Anterior spinocerebellar fasciculus
27. Ventral external arcuate fibers
28. Ventral median fissure
29. Spinothalamic fasciculus

THE INTERNAL PONS: CAUDAL SECTION

This discussion of the caudal section of the internal pons is not concerned with the ultimate connections of the various nuclei of the cranial nerves. These will be considered in the discussion on the cranial nerves, but you may find it useful to refer to the relevant diagrams while learning the features of the pons (see Figs. 63, 64, and 65).

The pons is continuous with the medulla, but it can be readily divided into a dorsal section, the **pars dorsalis,** or *pontine tegmentum,* and a ventral section, the **pars ventralis,** or *basilar pons.* It is easy to see that the tegmental portion of the pons is a continuation of the medulla. Many of the same tracts already presented in the sections on the medulla continue up into the dorsal pons. It is also the dorsal pons that contains the nuclei of the cranial nerves.

The ventral, or basilar, pons is made up of two basic kinds of fibers, the *longitudinal fasciculi* and the *transverse fibers.* The longitudinal fasciculi are composed of the long fibers of the **corticospinal tract** (22–24), which becomes the pyramid in the medulla, and of the shorter, **corticopontine fibers,** which originate in the cortex and terminate in various **pontine nuclei** (23–24), or *nuclei pontis.*

The transverse fibers, **fibrae pontis** (21–24), generally cross the midline to form the massive bundle of fibers known as the **middle cerebellar peduncle** (14–24), or *brachium pontis.*

Figure 24 shows the caudal portion of the pons at the level of the **abducens nerve [VI]** (17–24) and the **facial nerve [VII]** (12–24). The **fourth ventricle** (2–24), although still quite broad, is beginning to narrow. Just lateral to the midline in the floor of the fourth ventricle is a rounded protuberance called the **facial colliculus** (3–24). It is produced by the **genu of the facial nerve [VII]** (6–24). Just lateral to the facial colliculus is a depression known as the **superior fovea** (4–24).

The **superior vestibular nucleus** (1–24) is located in the lateral extent of the floor of the fourth ventricle. It sends fibers to the cerebellum in the **vestibulocerebellar fasciculus** (5–24), which courses along the medial portion of the **inferior cerebellar peduncle** (7–24), or *restiform body.* The **spinal tract and nucleus of the trigeminal nerve [V]** (9–24) are just medial to the middle cerebellar peduncle, and the **nucleus of the facial nerve [VII]** (15–24) is just medial to that. The **lateral vestibular nucleus** (25–24) lies just lateral to the superior vestibular nucleus.

The **superior olivary nucleus** (16–24) is ventral to the facial nucleus. It forms connections with the **trapezoid body** (20–24), which is connected to both superior olivary nuclei. At its lateral extent the trapezoid body forms the **lateral lemniscus** (11–24), which proceeds in a rostral direction. These anatomical features are associated with audition.

The **medial longitudinal fasciculus** (18–24) is near the floor of the fourth ventricle, and the **tectospinal tract** (24–24) is just ventral to the medial longitudinal fasciculus. The **medial lemniscus** (20–24) is now intermixed with the trapezoid body. The **ventral spinocerebellar fasciculus** (8–24) and the **lateral spinothalamic tract** (13–24) have been shifted to a more medial location. The **central tegmental tract** (10–24) is found dorsomedial to the lateral lemniscus.

Figure 24

Transverse section of the caudal portion of the pons.

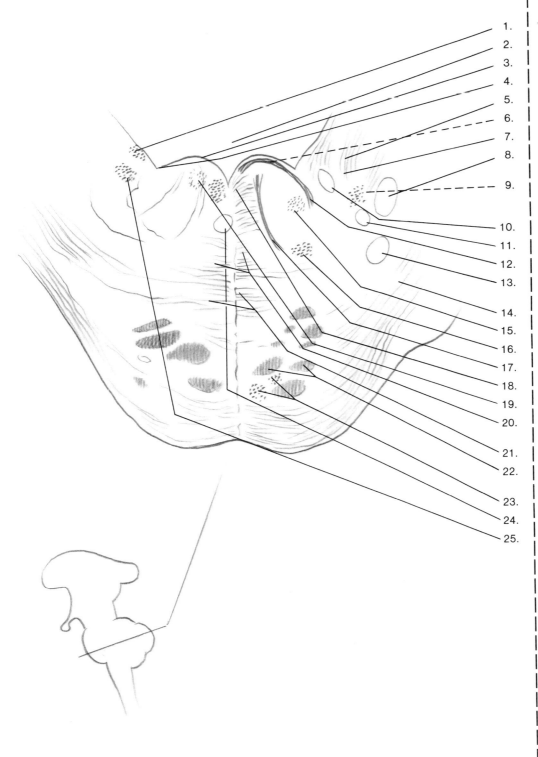

1.   *Superior vestibular nucleus
2.   Fourth ventricle
3.   Facial colliculus
4.   Superior fovea
5.   Vestibulocerebellar fasciculus
6.   Genu of facial nerve VII
7.   *Inferior cerebellar peduncle
8.   Ventral spinocerebellar fasciculus
9.   Spinal tract and nucleus of spinal tract of trigeminal nerve V
10. Central tegmental tract
11. Lateral lemniscus
12. Facial nerve VII
13. Lateral spinothalamic fasciculus
14. *Middle cerebellar peduncle
15. Nucleus of facial nerve VII
16. Superior olivary nucleus
17. Abducens nerve VI
18. Medial longitudinal fasciculus
19. Nucleus of abducens nerve VI
20. Trapezoid body and medial lemniscus
21. Fibrae pontis
22. Corticospinal and corticopontine tracts
23. *Pontine nuclei
24. Tectospinal fasciculus
25. *Lateral vestibular nuclei

THE INTERNAL PONS: MIDDLE SECTION

Figure 25 represents a section of the pons at the level of the **trigeminal nerve [V]** (17–25), which can be seen in the lateral portion of the figure proceeding in a ventro-lateral direction to emerge from the lateral aspect of the pons. The principal **sensory nucleus of the trigeminal nerve [V]** (9–25) is located at the dorsal extremity of the nerve, and the **motor nucleus of the trigeminal nerve [V]** (12–25) is ventrolateral to it. The **mesencephalic tract of the trigeminal nerve [V]** (7–25) is now apparent dorsal to the sensory and motor nuclei. The three cerebellar peduncles are all visible in the middle section. The **inferior cerebellar peduncle** (8–25), or *restiform body,* is now much smaller and has shifted to a position dorsal to the **middle cerebellar peduncle** (11–25), or *brachium pontis.* The **superior cerebellar peduncle** (3–25), or *brachium conjunctivum,* is located at the most dorsal aspect of this section.

The **fourth ventricle** (4–25), shown here enclosed by the **medullary velum** (1–25), is becoming more narrow. The **ventral spinocerebellar fasciculus** (2–25) has shifted to an even more dorsal position. The **facial colliculus** (5–25) can still be seen on the floor of the fourth ventricle, and the **genu of the facial nerve [VII]** (6–25) lies just below it. The **medial longitudinal fasciculus** (10–25) lies medial to the genu of the facial nerve, and the **tecto-spinal tract** (13–25) is just ventral to that. The **trapezoid body** (18–25) and the **medial lemniscus** (20–25) intermingled with it are in the same relative positions as in the caudal section. The **lateral lemniscus** (21–25) is the rostral projection of the trapezoid body and is lateral to the **superior olivary nucleus** (16–25).

The **corticospinal, corticobulbar,** and **corticopontine tracts** (24–25) are distributed throughout the basilar portion of the pons among the **pontine nuclei** (22–25). The reticular formation, which runs through the pons, is evident in this section, and the **oral pontine reticular nucleus** (14–25) is located in the medial tegmental area. The **lateral spinothalamic fasciculus** (15–25) is located just ventromedial to the trigeminal nerve.

The **aberrant pyramidal fibers** (23–25) are derived from the pyramidal system and enter the reticular system.

Figure 25

Transverse section of the pons at the level of the trigeminal-nerve root.

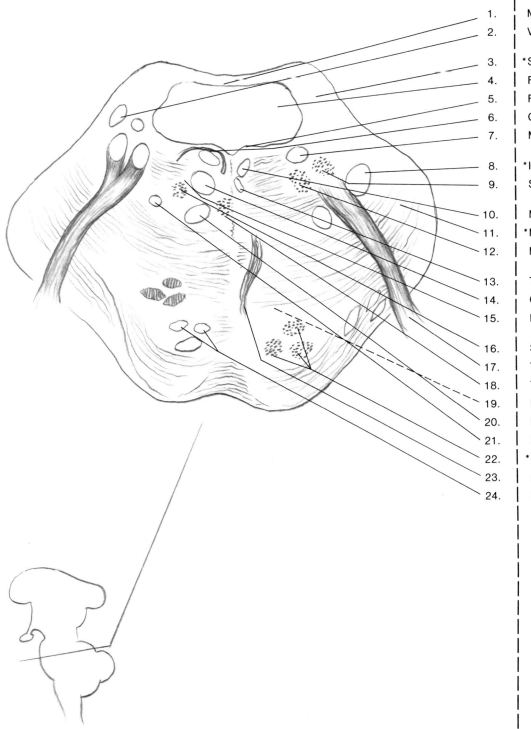

1.   Medullary velum
2.   Ventral spinocerebellar fasciculus
3.   *Superior cerebellar peduncle
4.   Fourth ventricle
5.   Facial colliculus
6.   Genu of facial nerve VII
7.   Mesencephalic tract of trigeminal nerve V
8.   *Inferior cerebellar peduncle
9.   Sensory nucleus of trigeminal nerve V
10.   Medial longitudinal fasciculus
11.   *Middle cerebellar peduncle
12.   Motor nucleus of trigeminal nerve V
13.   Tectospinal fasciculus
14.   Oral pontine reticular nucleus
15.   Lateral spinothalamic fasciculus
16.   Superior olivary nucleus
17.   Trigeminal nerve V
18.   Trapezoid and nuclei
19.   Fibrae pontis
20.   Medial lemniscus
21.   Lateral lemniscus
22.   *Pontine nuclei
23.   Aberrant pyramidal fibers
24.   Corticobulbar, corticospinal, and corticopontine fasciculi

THE INTERNAL PONS: ROSTRAL SECTION

In the rostral section of the internal pons (Fig. 26) the **corticospinal, corticobulbar,** and **corticopontine tracts** (22–26) can still be seen coursing through the **pontine nuclei** (20–26) in the basilar section of the pons. The **fourth ventricle** (5–26) is becoming more narrow and will soon become the **cerebral aqueduct** (7–27). The **anterior medullary velum** (4–26) encloses the roof of the fourth ventricle and at this point contains the **decussation of the trochlear nerve [IV]** (2–26). The **trochlear nerve [IV]** (1–26) itself can be seen exiting from the anterior medullary velum.

The **superior cerebellar peduncles** (18–26) lie on either side of the fourth ventricle near the lateral border. The **ventral spinocerebellar tract** (10–26), now relatively small, has shifted to a more dorsal position and is just lateral to the superior cerebellar peduncles. The **lateral lemniscus** (3–26) is ventral to the ventral spinocerebellar tract, and the **nuclei of the lateral lemniscus** (11–26) can also be seen in this section. The **medial lemniscus** (21–26) has now shifted to a somewhat more lateral position, but the **medial longitudinal fasciculus** (13–26) and the **tectospinal tract** (17–26) still maintain their relative positions just lateral to the midline. The **dorsal nucleus of raphe** (9–26), which receives some fibers from the superior peduncle, is evident just below the floor of the fourth ventricle. The **nucleus locus ceruleus** (8–26), or *pigmented nucleus,* the function of which is not really known but may be related to respiration and to sleep, is lateral to the nucleus of raphe; some of its cells are intermixed with the **mesencephalic nucleus of the trigeminal nerve [V]** (6–26).

The **central tegmental fasciculus** (14–26) continues in its same relative position, as does the **oral pontine reticular nucleus** (19–26) and the **lateral spinothalamic fasciculus** (12–26). The **middle cerebellar peduncle** (16–26) will be found in the most lateral aspect of Figure 26.

The **superior central nucleus** (15–26), which sends fibers to the cerebellum by way of deep, transverse fibers of the pons, is found at the midline.

The **reticular formation** is located lateral to the superior peduncle, and a lateral portion of it called the **pedunculopontine tegmental nucleus** (7–26) lies outside the superior cerebellar peduncle.

Figure 26

Transverse section of the rostral pons at the level of the trochlear nerve.

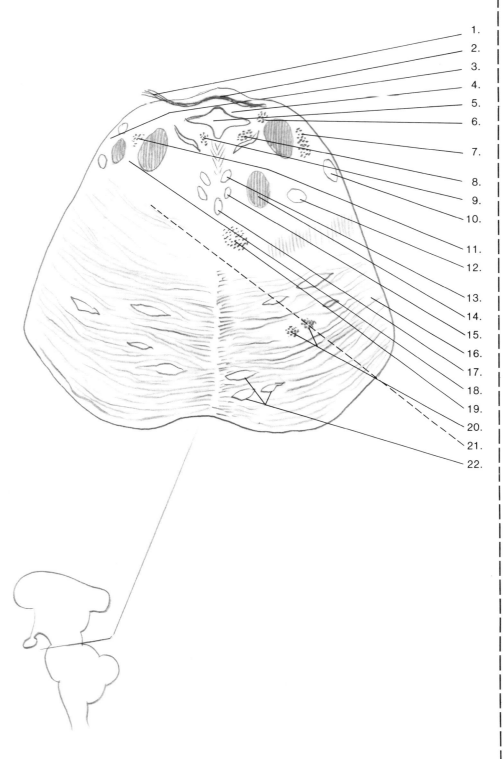

1. Trochlear nerve IV
2. Trochlear nerve decussation
3. Lateral lemniscus
4. Anterior medullary velum
5. Fourth ventricle
6. Mesencephalic nucleus of trigeminal nerve V
7. Pedunculopontine tegmental nucleus
8. *Nucleus locus ceruleus
9. Dorsal nucleus of raphe
10. Ventral spinocerebellar fasciculus
11. Nuclei of lateral lemniscus
12. Lateral spinothalamic fasciculus
13. Medial longitudinal fasciculus
14. Central tegmental tract
15. Superior central nuclei
16. *Middle cerebellar peduncle
17. Tectospinal fasciculus
18. *Superior cerebellar peduncle
19. Oral pontine reticular nucleus
20. *Pontine nuclei
21. Medial lemniscus
22. Corticobulbar, corticospinal, and corticopontine fasciculi

THE INTERNAL MESENCEPHALON: INFERIOR COLLICULUS

The **midbrain,** or **mesencephalon,** is a relatively short segment of the brain stem connecting the pons with the forebrain. The dorsal portion above the **cerebral aqueduct** (7–27) is variously called the **tectum** (1–27), the *corpora quadrigemina, quadrigeminal plate,* or *quadrigeminal bodies.* The **cerebral aqueduct** is also called the *aqueduct of Sylvius.* The tectum is made up of the **inferior colliculi** (4–27), or *postgemina,* and the **superior colliculi** (1–28), or *pregemina.*

That portion of the midbrain ventral to the tectum is called the **cerebral peduncle.** It is not a peduncle in the usual sense of a large bundle of nerve fibers since it consists of two major parts. The dorsal portion is the **midbrain tegmentum** (28–27), a direct continuation of the tegmentum of the pons. The ventral portion of the peduncle is the **cerebral peduncle** (15–27), or *basis pedunculi,* also called the *crus cerebri.* These two parts of the peduncle are separated by a large crescent-shaped nuclear mass, the **substantia nigra** (22–27), which is involved in the extrapyramidal system (see Sect. 58).

Figure 27 represents a section through the midbrain at the level of the **inferior colliculi** (4–27). The **nucleus of the inferior colliculus** (6–27), which is associated with auditory reflexes, receives most of the fibers of the **lateral lemniscus** (9–27) on the same side. Some of the lemniscal fibers course around the nucleus of the inferior colliculus through the **commissure of the inferior colliculus** (2–27) and terminate in the colliculus on the opposite side. Just lateral to the nucleus of the inferior colliculus is the **inferior quadrigeminal brachium** (5–27), which contains fibers from the lateral lemniscus that terminate in the **medial geniculate body** (9–28).

The **cerebral aqueduct** (7–27) is about 15 millimeters long and varies in shape from one section to another. It connects the fourth and the third ventricles. The aqueduct is surrounded by the **central gray stratum** (11–27), which at this level contains concentrations of cells that form the **nucleus of the trochlear nerve [IV]** (18–27) and the **nucleus of the mesencephalic root of the trigeminal nerve [V]** (8–27). The **nucleus locus ceruleus** (12–27), or *pigmented nucleus,* lies just below the mesencephalic-root nucleus. The **dorsal tegmental nucleus** (10–27), an expansion of the **dorsal nucleus of the raphe** (9–26), lies just above the trochlear nucleus. Just lateral to the central gray stratum the **pedunculopontine tegmental nucleus** (3–27) can be seen continuing up from the rostral pons. The **mesencephalic reticular formation** (25–27) is less extensive here than in the pons, but it still encom-

passes a good portion of the medial and lateral gray matter.

The **medial longitudinal fasciculus** (13–27) and the **tectospinal tract** (19–27) lie in the same relative positions as they did in the pons. The **superior central nucleus** (15–26) of the pons has now become the **ventral tegmental nucleus** (20–27).

The fibers of the **superior cerebellar peduncle** (21–27), or *brachium conjunctivum,* are deeply embedded in the tegmentum. They cross in the median plane at the **decussation of the superior cerebellar peduncle** (23–27) and turn rostrally; almost immediately most of them enter the **red nucleus** (19–28).

The **medial lemniscus** (27–27) continues up through the tegmentum of the mesencephalon and terminates in the thalamus. The **central tegmental tract** (16–27) also continues into the midbrain.

The tracts within the cerebral peduncle are organized according to their points of origin. Most laterally the fibers are derived from the cells in the temporal, parietal and occipital lobes and are called the **parieto-occipito-temporo-pontile tract** (14–27). The next, most medial collection of fibers is derived from the cells of the precentral gyrus in the cortex and is referred to as the **corticospinal tract** (17–27). The fibers of the **frontopontile tract** (24–27) originate in the cells of the frontal lobe and descend into the nuclei of the pons.

The depression between the cerebral peduncles is called the **interpenduncular fossa** (26–27), or *posterior perforated substance.* Just dorsal to the fossa lies the **interpeduncular nucleus** (24–28).

Figure 27

Transverse section through the mesencephalon at the level of the inferior colliculus.

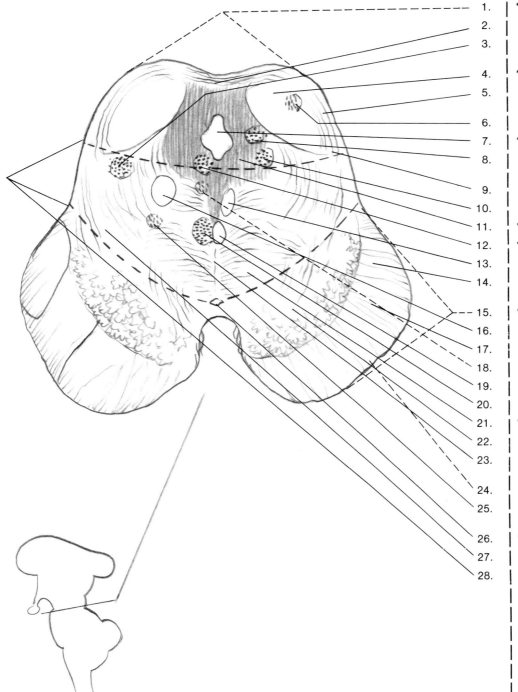

1. *Tectum
2. Commissure of inferior colliculi
3. Pedunculopontine tegmental nucleus
4. *Inferior colliculus
5. Inferior quadrigeminal brachium
6. Nucleus of inferior colliculus
7. *Cerebral aqueduct
8. Nucleus of mesencephalic root of trigeminal nerve V
9. Lateral lemniscus
10. Dorsal tegmental nucleus
11. *Central gray stratum
12. *Nucleus locus ceruleus
13. Medial longitudinal fasciculus
14. Parieto-occipito-temporo-pontile tract
15. *Cerebral peduncle
16. Central tegmental tract
17. Corticospinal fasciculus
18. Nucleus of trochlear nerve IV
19. Tectospinal fasciculus
20. Ventral tegmental nucleus
21. *Superior cerebellar peduncle
22. Substantia nigra
23. Decussation of superior cerebellar peduncle
24. Frontopontile tract
25. Mesencephalic reticular formation
26. *Interpeduncular fossa
27. Medial lemniscus
28. Midbrain tegmentum

THE INTERNAL MESENCEPHALON: SUPERIOR COLLICULUS

As can be seen in Figure 28 the most obvious feature of the midbrain at the level of the **superior colliculus** (1–28) is the **red nucleus** (19–28). It is oval in shape and extends rostrally as far as the posterior portion of the subthalamus. It derives its name from its pinkish color in fresh brain tissue. The red nucleus receives fibers from the frontal lobe of the cerebral cortex and is connected to the cerebellum through the superior cerebellar peduncle. It sends fibers to the spinal cord by way of the rubrospinal fasciculus.

The **dorsal tegmental decussation** (22–28), sometimes called the *fountain decussation of Meynert,* is made up of fibers connecting the superior colliculi with the tectobulbar and tectospinal fasciculi. The **ventral tegmental decussation** (23–28), or *decussation of Forel,* consists of fibers from the red nucleus crossing over the median raphe and descending as the rubrobulbar tract.

Just above the dorsal tegmental decussation and dorsomedial to the **medial longitudinal fasciculus** (15–28) lies the **nucleus of the oculomotor nerve [III].** Fibers of the **oculomotor nerve [III]** (26–28) exit from the ventral mesencephalon in the **interpeduncular fossa** (25–28), or *posterior perforated substance.*

The relative positions of the **substantia nigra** (18–28), the **cerebral peduncle** (28–28) and its component parts, the **parieto-occipito-temporo-pontile tract** (12–28), the **corticospinal tract** (16–28), the **frontopontile tract** (21–28), and the **pedunculopontine tract** (20–28) are essentially the same as at the level of the inferior colliculus.

Above the **central gray stratum** (6–28), also called the *periaqueductal gray,* which surrounds the **cerebral aqueduct** (8–28), the **commissure of the superior colliculus** (4–28) can be seen connecting the two colliculi. The **superior colliculus** (1–28) can be divided into four separate layers, or strata. The top stratum, composed of fine, white fibers, is the **stratum zonale** (2–28). The second layer, the **stratum griseum** (3–28), is thicker and contains a concentration of small, multipolar nerve cells embedded in a network of nerve fibers. The **stratum opticum** (5–28) and the **stratum lemnisci** (7–28) contain large numbers of myelinated fibers.

The **medial lemniscus** (27–28) can be seen in this section as crescent shaped. Dorsolateral to it lies a portion of the **inferior quadrigeminal brachium** (10–28), which flows into the **medial geniculate body** (9–28) from the **inferior colliculus** (4–27).

The **mesencephalic reticular formation** (11–28) is prominent in this section and is in the same relative position as at the level of the inferior colliculus, as is the **central tegmental tract** (14–28). The **dorsal tegmental nucleus** (13–28) lies just above the **medial longitudinal fasciculus** (15–28).

Figure 28

Transverse section through the mesencephalon at the level of the superior colliculus.

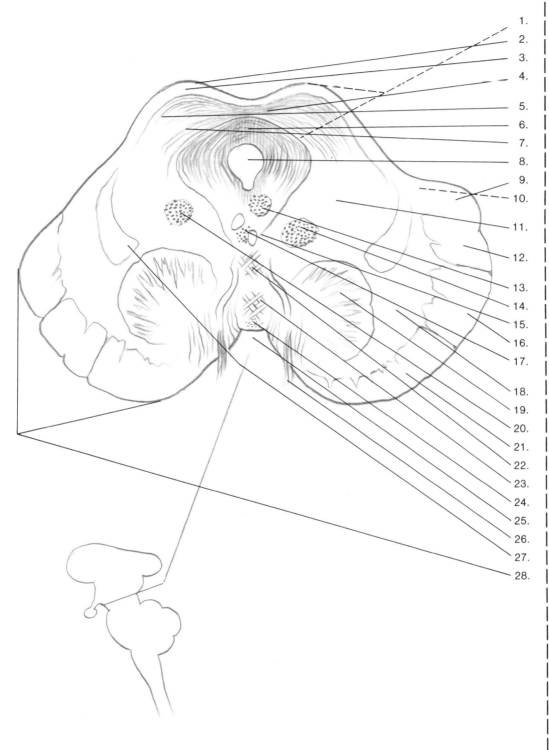

1. *Superior colliculus
2. Stratum zonale
3. Stratum griseum
4. Commissure of superior colliculus
5. Stratum opticum
6. *Central gray stratum
7. Stratum lemnisci
8. *Cerebral aqueduct
9. Medial geniculate body
10. Inferior quadrigeminal brachium
11. Mesencephalic reticular formation
12. Parieto-occipito-temporo-pontile tract
13. Dorsal tegmental nucleus
14. Central tegmental tract
15. Medial longitudinal fasciculus
16. Corticospinal fasciculus
17. Nucleus of oculomotor nerve III
18. Substantia nigra
19. Red nucleus
20. Pedunculopontine tract
21. Frontopontile tract
22. *Dorsal tegmental decussation
23. *Ventral tegmental decussation
24. Interpeduncular nucleus
25. *Interpeduncular fossa
26. Oculomotor nerve III
27. Medial lemniscus
28. *Cerebral peduncle

THE DIENCEPHALON: OVERVIEW

The diencephalon is one of the most complex portions of the neuroaxis, and most body functions are represented somewhere in its interconnections. The diencephalon lies between the midbrain and the cerebral hemispheres. It is divided into five major portions. These include the **subthalamus** (27–29), which is a continuation of the mesencephalic tegmentum; the **epithalamus** (8–29), which is the roof of this region and includes the **pineal body** (5–29), **habenular complex** (4–29), and **posterior commissure** (12–29); the **metathalamus** (32–29), which includes the **lateral geniculate body** (33–29) and the **medial geniculate body** (34–29); the **hypothalamus** (15–29); and the **thalamus** (6–29) proper, or *dorsal thalamus.* Figure 29 will help to clarify their relationships.

The **subthalamus** (27–29), or *ventral thalamus,* is the transition area between the tegmentum of the mesencephalon and the rest of the diencephalon. It contains the most rostral portions of the **red nucleus** (18–29) and the **substantia nigra** (26–29). The nuclear masses in the subthalamus include the **subthalamic nucleus** (16–29), located dorsolateral to the rostral portion of the substantia nigra. It appears to have connections with the red nucleus and the substantia nigra, as well as to other nuclei in the subthalamus. The *zona incerta* (not shown in this figure) lies lateral to the subthalamic nucleus and runs just ventral to the thalamus through most of the diencephalon. The **tegmental field of Forel** (17–29) lies medial to portions of the zona incerta and runs in front of the red nucleus. The cells in this tegmental field are the *nucleus of the field of Forel* and are a continuation of the reticular system of the mesencephalon. Its cells connect with the **lenticular nucleus** (3–9) through the *ansa lenticularis.*

The **epithalamus** (8–29) may be considered the roof of the diencephalon and is composed of the **habenular complex** (4–29), the **pineal body** (5–29), and the **posterior commissure** (12–29). The habenular complex lies just rostral to the pineal body and is a part of the olfactory correlating mechanism. It contains the lateral and medial habenular nuclei, which receive fibers from the **stria medullaris** (3–29), which projects along the dorsomedial surface of the thalamus making connections with the olfactory centers of the cerebral hemispheres. The pineal body or *epiphysis cerebri,* is an endocrine gland the functions of which are not yet clear. It receives some fibers from the habenula and from the posterior commissure, but they do not appear to have any functional significance. The pineal body lies on the dorsal aspect of the mesencephalon between the two thalami. The posterior commissure is a large bundle of fibers that crosses the midline at the point at which the **cerebral aqueduct** (28–29) begins to expand into the **third ventricle** (10–29). These fibers connect some of the cells in the two superior colliculi. Others connect the midbrain reticular substance on the two sides, and some are probably associated with the cells of the posterior thalami.

The **metathalamus** (32–29) is composed of the **medial geniculate bodies** (34–29) and the **lateral geniculate bodies** (33–29), one of each positioned on either side of the brain. The medial geniculate bodies are located just lateral to the colliculi (see 22–8). They have connections with the inferior colliculi through the **inferior quadrigeminal brachium** (10–28). The two geniculate bodies are connected through the **commissure of Gudden,** which crosses the midline in the posterior portion of the **optic chiasm** (19–29). The medial geniculate bodies project fibers to and receive fibers from the temporal lobe of the cortex and are associated with audition. The lateral geniculate bodies are associated with vision and are rounded eminences located on the posterolateral portion of the thalamus. They have connections with the superior colliculi and also project fibers to the occipital cortex in the area of the **calcarine fissure** (25–3).

The **hypothalamus** (15–29) (Figures 30–35) is the ventral portion of the diencephalon and forms the floor and part of the walls of the **third ventricle** (10–29) below the **hypothalamic sulcus** (14–29). It includes the **mammillary bodies** (21–29), the **tuber cinereum** (20–29), the optic chiasm, and the **infundibulum** (22–29). Ventrally, the hypothalamus extends from the optic chiasm to the posterior portion of the mammillary bodies. The subthalamus lies caudal and lateral to the hypothalamus, and the thalamus is dorsal to the subthalamus.

The *dorsal thalamus* is most frequently referred to simply as the **thalamus** (6–29). There are two thalami, which lie on either side of the third ventricle. They are joined by a flattened mass of gray matter consisting of the central medial nuclei and called the **intermediate mass** (11–29). The posterior portion of the thalamus, the **pulvinar** (2–29), extends caudally and covers the superior colliculi. The lateral geniculate body is located on the lateral aspect of the pulvinar, and the medial geniculate is beneath it. The thalamus is bounded laterally by the internal capsule; its dorsal surface is free. It is separated laterally from the **caudate nucleus** (1–9) by the **stria terminalis** (7–8).

Figure 29

Overview of the components of the diencephalon.

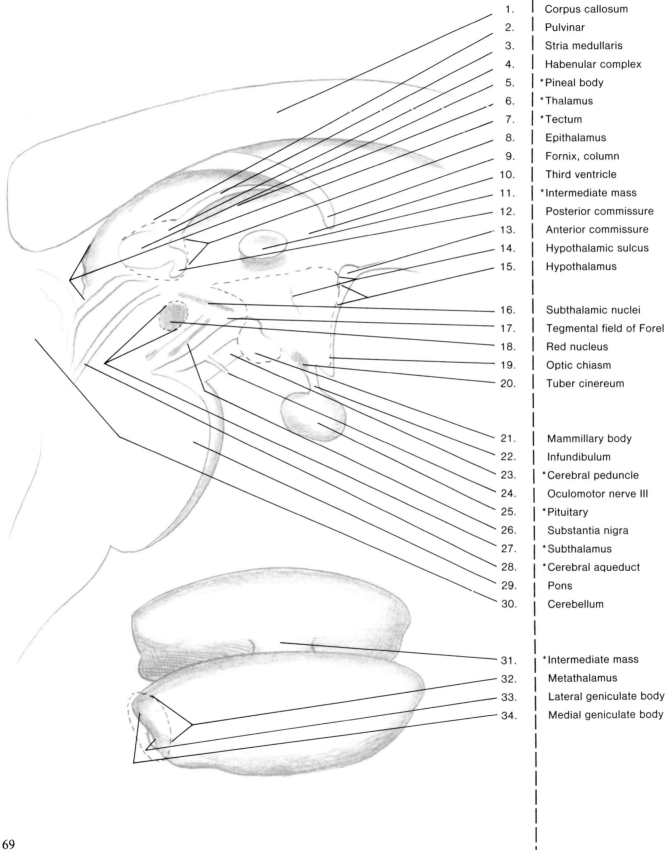

1. Corpus callosum
2. Pulvinar
3. Stria medullaris
4. Habenular complex
5. *Pineal body
6. *Thalamus
7. *Tectum
8. Epithalamus
9. Fornix, column
10. Third ventricle
11. *Intermediate mass
12. Posterior commissure
13. Anterior commissure
14. Hypothalamic sulcus
15. Hypothalamus

16. Subthalamic nuclei
17. Tegmental field of Forel
18. Red nucleus
19. Optic chiasm
20. Tuber cinereum

21. Mammillary body
22. Infundibulum
23. *Cerebral peduncle
24. Oculomotor nerve III
25. *Pituitary
26. Substantia nigra
27. *Subthalamus
28. *Cerebral aqueduct
29. Pons
30. Cerebellum

31. *Intermediate mass
32. Metathalamus
33. Lateral geniculate body
34. Medial geniculate body

THE HYPOTHALAMIC NUCLEI: THREE-DIMENSIONAL VIEW

Although the hypothalamus is a very small portion of the brain, it is involved directly or indirectly in a large number of critical life functions. These will be covered in more detail in later sections on functional systems, but it should be noted here that they include the regulation of sleep, food consumption, water consumption, temperature regulation, a variety of emotional expressions, and a large number of sympathetic and parasympathetic functions.

The general boundaries of the hypothalamus were described in Section 27. Although differentiation of the nuclei within the hypothalamus is far from precise, it is possible to identify specific nuclei. Some of these are more readily identified in lower animal forms, and some are more distinct in the fetus than they are in the adult. It is also true that within a given nucleus there may be several kinds of cells that can be differentiated histologically. The localization of function, with some exceptions, is not specific to the individual nuclei, but tends to overlap the nuclear boundaries; thus function is related to hypothalamic regions. However, the nuclei provide landmarks that are useful in understanding discussions of function. Therefore a good three-dimensional map of this important part of the brain is valuable. Figure 30 presents an overview of the nuclei, and Figures 31–33 present a series of frontal sections. Both types of representation are essential for a good understanding of the area. It must be emphasized that these drawings are schematics for teaching purposes. The actual nuclei are not well differentiated and neatly separated into little structures.

Histological study of the hypothalamus indicates that it can be reasonably divided into three general areas around the rostral-caudal axis. These are the *periventricular zone*, the *medial zone*, and the *lateral zone*. The periventricular zone, next to the third ventricle, consists of an extension of the **central gray stratum** (6–28) of the midbrain, which surrounds the **cerebral aqueduct** (8–28). The periventricular zone contains the **periventricular nucleus** (9–31), the ventral portion of which in the region of the **tuber cinereum** (20–29) can be differentiated from the rest of the nucleus and is given the name **arcuate nucleus** (9–32). The medial zone lies outside the periventricular zone, and the lateral hypothalamic zone consists of relatively undifferentiated masses of cells and fibers that run along the most lateral aspect of the hypothalamus. All three zones extend through the entire length of the hypothalamus.

It is also useful to divide the hypothalamus into three parts transversally. These can be recognized as the rostral or **supraoptic region** (17–30), the middle or **tuberal region** (20–29), and the most caudal or **mammillary region** (15–30). The preoptic area lies just above the **optic chiasm** (20–30) and is generally considered a part of the telencephalon, although it is closely related to the hypothalamic structures. It contains the **medial preoptic nucleus** (18–30) and the **lateral preoptic nucleus** (13–30). The **paraventricular nucleus** (7–30) and the **supraoptic nucleus** (17–30), unlike most of the nuclei of the hypothalamus, are quite sharply defined. Both lie in the supraoptic region. The supraoptic nucleus is located directly over the lateral portion of the optic chiasm. The paraventricular nucleus extends dorsally almost to the hypothalamic sulcus and rostrally nearly to the optic chiasm. The **anterior nucleus** (12–30), which is not nearly as well defined, lies above the supraoptic nucleus and tends to merge with the preoptic nuclei.

The hypothalamus is widest in the middle, or tuberal region and is separated from the lateral hypothalamic areas by the fibers of the **fornix** (2–30). The **ventromedial nucleus** (16–30) and the **dorsomedial nucleus** (11–30) tend to blend with one another and are both found in the tuberal area. The **dorsal hypothalamic area** (4–32) lies just above the dorsomedial nucleus. The **posterior nucleus** (10–30) is usually considered to belong in the tuberal region, although the more caudal portion is in the mammillary section.

The most prominent feature of the mammillary region is the large mammillary body, which consists of the **medial mammillary nucleus** (17–33), the **intermediate mammillary nucleus** (14–33), also called the *intercalated nucleus,* and the **lateral mammillary nucleus** (16–33).

Figure 30

Three-dimensional view of the nuclei of the hypothalamus.

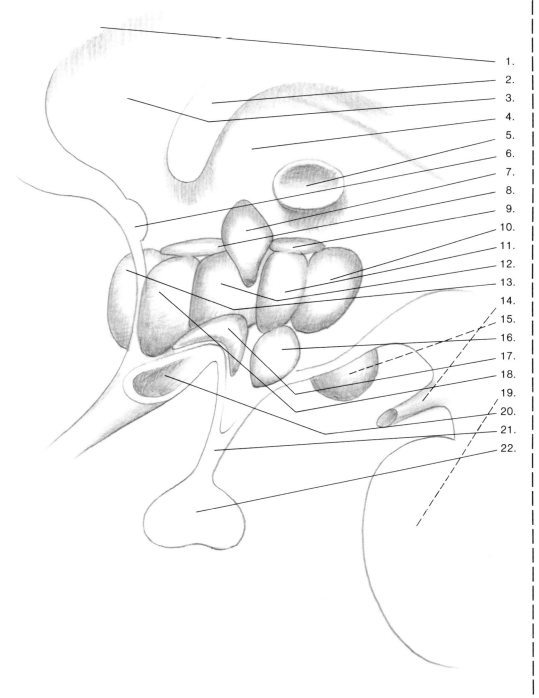

1. Corpus callosum
2. Fornix
3. Septum pellucidum
4. *Thalamus
5. *Intermediate mass
6. Anterior commissure
7. Paraventricular nucleus
8. Lateral area
9. Dorsal area
10. Posterior nucleus
11. Dorsomedial nucleus
12. Anterior nucleus
13. Lateral preoptic nucleus
14. Oculomotor nerve III
15. Mammillary body
16. Ventromedial nucleus
17. Supraoptic nucleus
18. Medial preoptic nucleus
19. Pons
20. Optic chiasm
21. Infundibulum
22. *Pituitary

THE INTERNAL HYPOTHALAMUS: SUPRAOPTIC REGION

The coronal section of the brain in Figure 31 shows the anterior hypothalamus just behind the preoptic nuclei. Other structures are also labeled to provide a general orientation. The reasonably well-defined **paraventricular nucleus** (10–31) and **supraoptic nucleus** (14–31) can be readily seen. The cellular structure of these two nuclei is similar, and there is some evidence that the cells have secretory activity. Both nuclei send fibers to the posterior lobe of the pituitary. The two supraoptic nuclei are connected by a few fibers that cross in the bridge of tissue over the **optic chiasm** (15–31).

This transverse section shows clearly how closely the **periventricular nucleus** (9–31) lies around the **third ventricle** (11–31). The **lateral hypothalamic area** (12–31), which merges rostrally with the **lateral preoptic nucleus** (13–30), runs throughout the length of the hypothalamus, as can be noted in Figures 31–33. Much of this area is composed of fibers of the **medial forebrain bundle** (1–34(d)). The **anterior nucleus** (13–31), which is not particularly well defined, blends with the lateral area to the side and with the **medial preoptic nucleus** (18–30) anteriorly. This hypothalamic area is associated with the regulation of body temperature.

The supraoptic region of the hypothalamus has rich connections with the dorsal thalamus and with the cerebral hemispheres. It also has important connections with the **septum pellucidum** (2–31).

Figure 31

Transverse section of the hypothalamus through the supraoptic region.

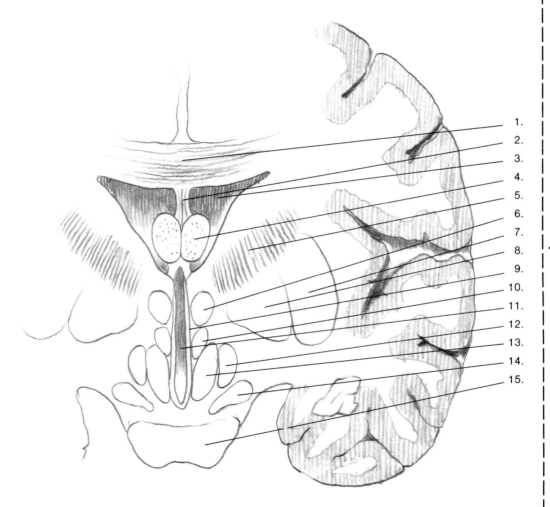

1. Corpus callosum
2. Septum pellucidum
3. Lateral ventricle
4. Fornix, body
5. Internal capsule
6. Fornix, column
7. Putamen
8. *Globus pallidus
9. Periventricular nucleus
10. Paraventricular nucleus
11. Third ventricle
12. Lateral hypothalamic area
13. Anterior nucleus
14. Supraoptic nucleus
15. Optic chiasm

THE INTERNAL HYPOTHALAMUS: TUBERAL REGION

In the tuberal region of the hypothalamus (Fig. 32) the **lateral hypothalamic area** (8–32) continues in a caudal direction and is now separated from the other hypothalamic nuclei by the fibers of the **fornix** (1,7–32). The **ventromedial nucleus** (13–32) and the **dorsomedial nucleus** (6–32) are both found in this region, but are not clearly distinguished from each other. Both of these nuclei are associated with emotional expression, and the ventromedial nucleus has been shown to produce obesity in some animals when lesioned. The **dorsal hypothalamic area** (4–32) is located just above the dorsomedial nucleus.

The ventral portion of the **periventricular nucleus** (5–32) has broadened and thickened to form the **arcuate nucleus** (9–32). The **optic tract** (11–32) has separated from the **optic chiasm** (15–31) and is located lateral to the hypothalamus, forming its lateral boundary. A small portion of the **supraoptic nucleus** (10–32) can be seen just medial to the optic tract. The **nucleus tuberis** (12–32) is located in the ventral portion of the lateral area. These nuclei may form enlargements on the surface of the **tuber cinereum** (14–32). They send fibers to the hypophysis and are believed to be a part of the neuroendocrine system.

Two of the hypothalamic fiber bundles are represented in Figure 32. The **pallidohypothalamic tract** (15–32) runs between the **globus pallidus** (16–32) and the ventromedial nucleus. There is also evidence that the ventromedial nucleus has connections with the frontal lobe and with the amygdala. The **mammillothalamic tract** (2–32), which runs between the **anterior thalamic nuclei** (11–38) and the **mammillary body** (15–30), is shown in the center of the **thalamus** (4–30).

Figure 32

Transverse section of the hypothalamus through the tuberal region.

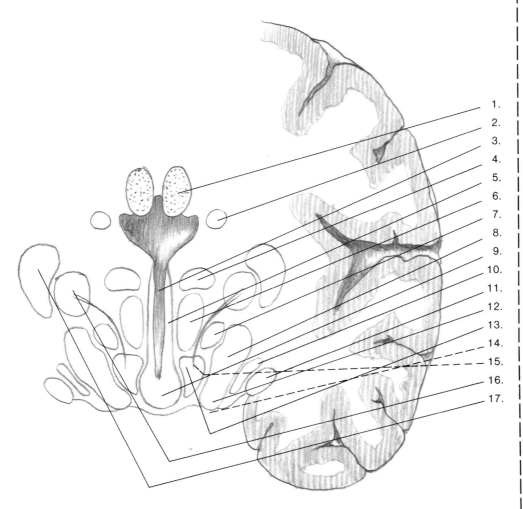

1.   Fornix, body
2.   Mammillothalamic tract
3.   Third ventricle
4.   Dorsal hypothalamic area
5.   Periventricular nucleus
6.   Dorsomedial nucleus
7.   Fornix, column
8.   Lateral hypothalamic area
9.   Arcuate nucleus
10.   Supraoptic nucleus
11.   Optic tract
12.   Nucleus tuberis
13.   Ventromedial nucleus
14.   Tuber cinereum
15.   Pallidohypothalamic tract
16.   *Globus pallidus
17.   Putamen

THE INTERNAL HYPOTHALAMUS: MAMMILLARY REGION

The most characteristic feature of the mammillary region of the internal hypothalamus (Fig. 33) is, of course, the mammillary body, which is appropriately shaped. The **medial mammillary nucleus** (17–33) is much larger than the **lateral mammillary nucleus** (16–33). There are histological differences in the medial and lateral portions of the medial mammillary nucleus; the cells of the medial portion are relatively larger than those of the more lateral part. The **intermediate mammillary nucleus** (14–33), or *intercalated nucleus,* also considered part of the mammillary body, is located just dorial to the lateral mammillary nucleus. The mammillary body, as will be indicated in more detail in a later section (see Sects. 32 and 33), has extensive connections with other parts of the brain. The **fornix** (2–33), which connects the **hippocampus** (32–45) with the lateral portion of the medial mammillary nucleus and other parts of the mammillary formation can be seen above the thalamus.

Dorsal to the mammillary structure is the relatively large **posterior nucleus** (8–33). It extends dorsally up to the hypothalamic sulcus. In a rostral direction it extends into the tuberal region to the **dorsal hypothalamic area** (4–32) and the **dorsomedial nucleus** (6–32).

The relationship between this, the most caudal, portion of the hypothalamus and the subthalamus can be seen in Figures 33 and 34. This cross section shows that the **subthalamic nucleus** (15–33) and the **cerebral peduncle** (11–33) are lateral to the most caudal portion of the **lateral hypothalamic area** (12–33).

Figure 33

Transverse section of the hypothalamus through the mammillary region.

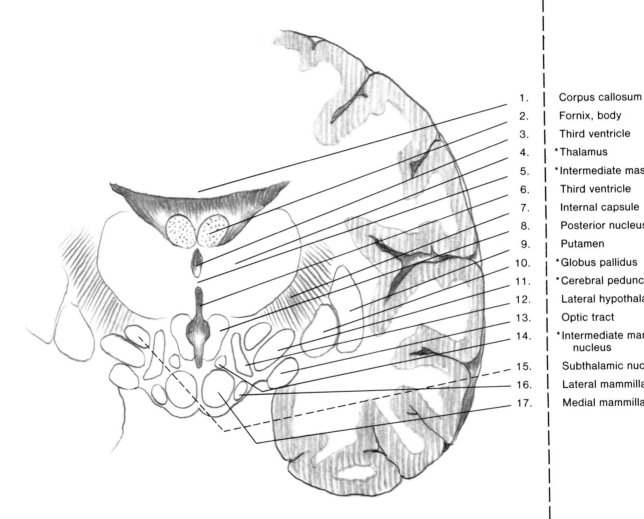

1. Corpus callosum
2. Fornix, body
3. Third ventricle
4. *Thalamus
5. *Intermediate mass
6. Third ventricle
7. Internal capsule
8. Posterior nucleus
9. Putamen
10. *Globus pallidus
11. *Cerebral peduncle
12. Lateral hypothalamic area
13. Optic tract
14. *Intermediate mammillary nucleus
15. Subthalamic nucleus
16. Lateral mammillary nucleus
17. Medial mammillary nucleus

HYPOTHALAMIC CONNECTIONS: AFFERENT

The fiber connections within the hypothalamus and between the hypothalamus and the rest of the central nervous system are too complex to be represented in their entirety in two drawings. However, Figures 34 and 35 graphically present the major afferent and efferent connections. Learning these principal fiber tracts and their points of origin and termination will greatly facilitate the understanding of functional relationships, which will be covered in later sections.

The **medial forebrain bundle** (1–34d) is a significant group of fibers, more prominent in lower animals than in man. It originates in the **basal olfactory regions** and the **septum** (2–34d). The bundle contains both ascending and descending neurons. It courses through the **lateral hypothalamic area** (4–34b) into the **midbrain tegmentum** (7–34d), making connections with the **preoptic nucleus** (5–34d) and the **ventromedial nucleus** (6–34d) of the hypothalamus. It is believed that the medial forebrain bundle carries impulses that affect, in particular, the regulation of the hypothalamic sympathetic centers.

The **fornix** (1–34a) is a very large and prominent bundle of fibers that originates in the large pyramidal cells of the hippocampus and ends primarily in the **medial mammillary nucleus** (5–34a). It does send fibers to the **lateral preoptic nuclei** (3–34a) and to the **dorsal hypothalamic nucleus** (4–34a), however. Some fibers also go to the **habenular complex** (6–34a) of the epithalamus.

The **stria terminalis** (2–34c) has its origin in the most caudal portion of the **amygdaloid complex** (6–40). It travels along the **caudate nucleus** (1–40) and terminates in the **medial preoptic nucleus** (3–34c) and the **anterior hypothalamic nucleus** (4–34c), with some fibers going to the ventromedial nucleus.

The **inferior thalamic peduncle** (1–34b) begins in the **dorsomedial nucleus of the thalamus** (2–34b) and the **midline thalamic nuclei** (3–34b). It sends fibers to the **lateral preoptic nuclei** (3–34a) and the **lateral hypothalamic area** (4–34b).

Finally, the **mammillary peduncle** (5–34b) originates in the **ventral tegmental nucleus** (8–34b) and the **dorsal tegmental nucleus** (7–34b) of the midbrain, projecting fibers to the **lateral mammillary nucleus** (6–34b).

Figure 34

Afferent pathways to the hypothalamus.

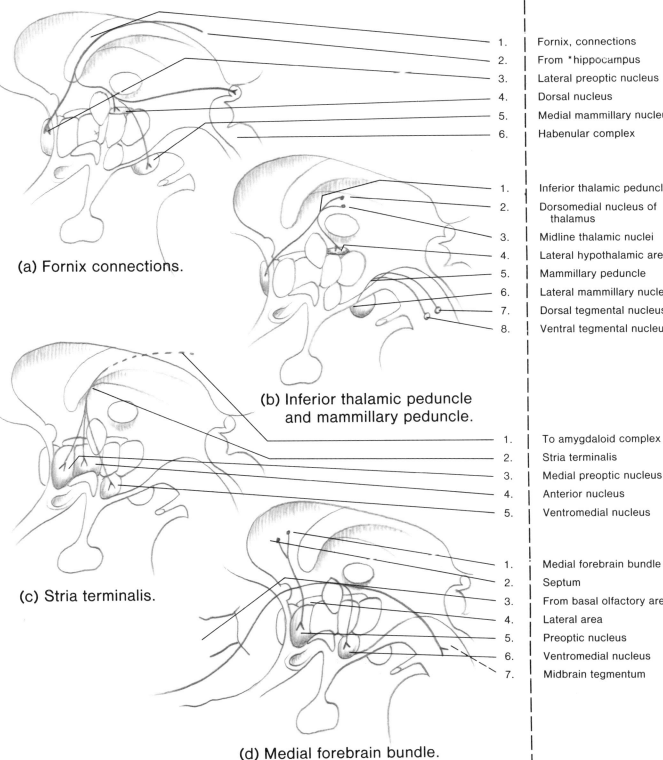

(a) Fornix connections.

1. Fornix, connections
2. From *hippocampus
3. Lateral preoptic nucleus
4. Dorsal nucleus
5. Medial mammillary nucleus
6. Habenular complex

(b) Inferior thalamic peduncle and mammillary peduncle.

1. Inferior thalamic peduncle
2. Dorsomedial nucleus of thalamus
3. Midline thalamic nuclei
4. Lateral hypothalamic area
5. Mammillary peduncle
6. Lateral mammillary nucleus
7. Dorsal tegmental nucleus
8. Ventral tegmental nucleus

(c) Stria terminalis.

1. To amygdaloid complex
2. Stria terminalis
3. Medial preoptic nucleus
4. Anterior nucleus
5. Ventromedial nucleus

(d) Medial forebrain bundle.

1. Medial forebrain bundle
2. Septum
3. From basal olfactory area
4. Lateral area
5. Preoptic nucleus
6. Ventromedial nucleus
7. Midbrain tegmentum

HYPOTHALAMIC CONNECTIONS: EFFERENT

The hypothalamus has important efferent connections (Fig. 35) with the **posterior lobe of the pituitary** (6–35a), also called the *neuropituitary,* which contributes to the regulation of some of the pituitary hormones. The **supraopticohypophysial tract** (3–35a) is a well-defined bundle of fibers that comes principally from the **supraoptic nucleus** (2–35a) and the **paraventricular nucleus** (1–35a). A smaller tract, the **tuberohypophysial tract** (5–35a), connects the medial cells of the **tuber cinereum** (14–32) with the posterior lobe of the hypophysis.

The **principal mammillary fasciculus** (4–35c) emerges from the **medial mammillary nucleus** (5–34a) with contributions from the **lateral mammillary nucleus** (6–34b) and the **intermediate mammillary nucleus** (14–33). It proceeds in a dorsal direction for a short distance and divides into the **mammillothalamic tract** (3–35c) and the **mammillotegmental tract** (5–35c). The mammillothalamic tract contains fibers from the medial mammillary nucleus that proceed to the thalamus, ending in the **anteroventral nucleus** (2–35c) and the **anteromedial nucleus** (1–35c). Those fibers in the tract that originate in the lateral mammillary nucleus project to the anterodorsal nucleus.

The mammillotegmental tract projects to the midbrain tegmentum and terminates in the **dorsal tegmental nucleus** (8–35c) and the **ventral tegmental nucleus** (7–35c). It also sends fibers to the reticular system of the brain stem.

The **periventricular system** (1–35b) is a significant bundle of fibers that projects in both anterior and posterior directions. Those fibers running in an anterior direction proceed from the **posterior nucleus** (8–33), with some contributions from the supraoptic nucleus and the tuberal nucleus, to the **dorsomedial nucleus of the thalamus** (2–35b) and the **midline nuclei of the thalamus** (3–35b). Some fibers from these nuclei originate in the thalamus and run to the hypothalamus. Most of the fibers of the periventricular system turn in a caudal direction and run through the **dorsal longitudinal fasciculus** (5–35b) to the ventral portion of the **central gray** (8–35b) and the **dorsal tegmental nucleus** (9–35b). Much of this descending branch carries efferent autonomic fibers to nuclei of both the sympathetic and parasympathetic systems. A few of the fibers from the caudal portion of the periventricular system end in the **midbrain tectum** (7–35b).

Figure 35

Efferent pathways from the hypothalamus.

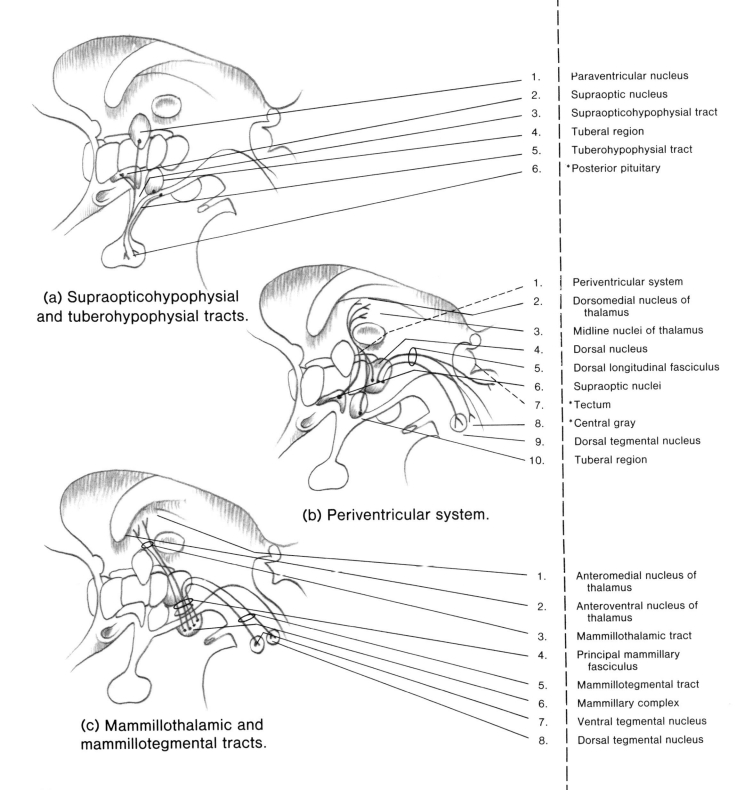

(a) Supraopticohypophysial and tuberohypophysial tracts.

1. Paraventricular nucleus
2. Supraoptic nucleus
3. Supraopticohypophysial tract
4. Tuberal region
5. Tuberohypophysial tract
6. *Posterior pituitary

(b) Periventricular system.

1. Periventricular system
2. Dorsomedial nucleus of thalamus
3. Midline nuclei of thalamus
4. Dorsal nucleus
5. Dorsal longitudinal fasciculus
6. Supraoptic nuclei
7. *Tectum
8. *Central gray
9. Dorsal tegmental nucleus
10. Tuberal region

(c) Mammillothalamic and mammillotegmental tracts.

1. Anteromedial nucleus of thalamus
2. Anteroventral nucleus of thalamus
3. Mammillothalamic tract
4. Principal mammillary fasciculus
5. Mammillotegmental tract
6. Mammillary complex
7. Ventral tegmental nucleus
8. Dorsal tegmental nucleus

THE NUCLEI OF THE THALAMUS

Sensory impulses of all kinds from specific and general receptors converge on the thalamus and metathalamus. This relatively complex nuclear mass functions as a relay station, receiving the afferent fibers and, after interruption, projecting them to the cortex. Sections 34, 35, and 36 will deal with this portion of the diencephalon. Figure 36 shows a three-dimensional view of the various nuclei. Some of the nuclei that are buried can best be visualized in cross section, as in Figure 37. This figure also shows the relationship between the thalamus and other structures. Frequent reference to Figures 36 and 37 will facilitate study of the thalamus. The text of this section will provide the label numbers for both figures. Figure 38 shows the principal inputs to the various nuclei and their projections to the cortex. Figure 39 shows the areas of the cortex to which the thalamic nuclei project.

The thalamus is a large ovoid mass composed primarily of gray matter, although it has some significant portions that are white. One of those is a thin layer of fibers that covers the dorsal surface and that is known as the **stratum zonale.** The thalamus is bounded laterally by the **internal capsule** (7–37). The most lateral nucleus of the thalamus is relatively narrow and is called the **reticular nucleus** (10–37). This small nuclear mass is separated from the rest of the thalamus by a broad band of myelinated fibers, the **external medullary lamina** (6–37).

The thalamus is divided into lateral and medial parts by another band of myelinated fibers called the **internal medullary lamina** (10–36), (9–37). The anterior portion of the internal medullary lamina divides into two sections, defining, in part, the **anterior nuclear group of the thalamus** (15–36), (3–37). The anterior portion of the thalamus is somewhat raised, forming the anterior tubercle. The anterior nuclear group is composed of three identifiable nuclei, the *anteroventral, anterodorsal,* and *anteromedial nuclei.*

The medial portion of the thalamus is divided into the relatively large **dorsomedial nucleus** (5–36), (8–37) and the smaller, more medial **midline nucleus** (2–36). The midline nucleus extends across the third ventricle in the **intermediate mass** (1–36), also called the *interthalamic adhesion.* The intermediate mass is reported to be absent in about 30 percent of the adult human population.

Several nuclei infiltrate the internal medullary lamina and separate the medial nuclei from the lateral. The largest and most important of these intralaminar nuclei is the **centromedian nucleus** (12–36), (11–37). It is lo-cated in the middle third of the thalamus and is almost completely surrounded by the fibers of the internal medullary lamina. The rest of the intralaminar nuclei are small and have indistinct boundaries.

The lateral nuclei of the thalamus are divided into two general groups, the smaller, dorsal group and the larger, ventral group. The **lateral dorsal nucleus** (9–36) is centrally located in the dorsal thalamus and is considered by some anatomists to be a posterior extension of the anterior nuclear group. The **lateral posterior nucleus** (4–36) is located just posterior to the lateral dorsal nucleus and dorsal to the **ventral posterior nucleus** (7,11–36). Caudally it is bounded by the **pulvinar** (3–36), the large, posterior portion of the thalamus that overhangs the **geniculate bodies** (6,8–36) and the **superior colliculi** (1–28).

The ventral nuclear mass of the thalamus is generally divided into three separate nuclei: the **ventral anterior nucleus** (14–36), the **ventral lateral nucleus** (13–36), and the ventral posterior nucleus. The last is divided further into the **ventral posterior lateral nucleus** (7–36) and the **ventral posterior medial nucleus** (11–36), (18–37). The ventral posterior nucleus is also called the *semilunar nucleus* or *arcuate nucleus.*

Figure 36

Three-dimensional view of the nuclei of the thalamus.

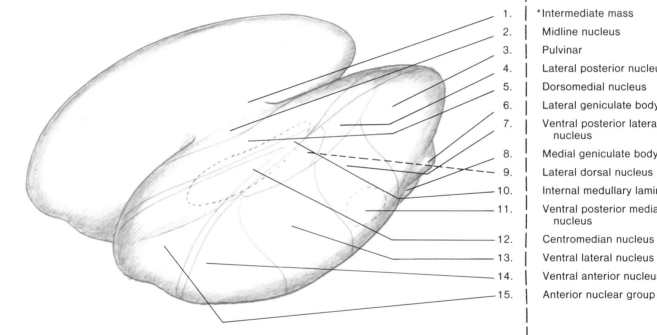

1. *Intermediate mass
2. Midline nucleus
3. Pulvinar
4. Lateral posterior nucleus
5. Dorsomedial nucleus
6. Lateral geniculate body
7. Ventral posterior lateral nucleus
8. Medial geniculate body
9. Lateral dorsal nucleus
10. Internal medullary lamina
11. Ventral posterior medial nucleus
12. Centromedian nucleus
13. Ventral lateral nucleus
14. Ventral anterior nucleus
15. Anterior nuclear group

Figure 37

Frontal section through the thalamus and the structures in its immediate vicinity.

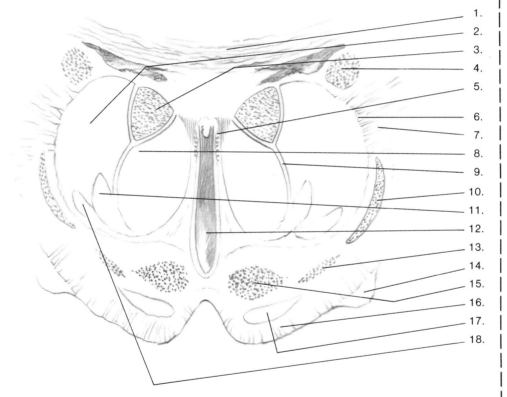

1. Corpus callosum
2. Lateral portion of thalamus
3. Anterior nuclear group
4. Caudate nucleus
5. Choroid plexus of third ventricle
6. External medullary lamina
7. Internal capsule
8. Dorsomedial nucleus
9. Internal medullary lamina
10. Reticular nuclei
11. Centromedian nucleus
12. Third ventricle
13. Subthalamic nucleus
14. Lateral geniculate body
15. Red nucleus
16. *Cerebral peduncle
17. Substantia nigra
18. Ventral posterior medial nucleus

THE THALAMUS: INPUT AND PROJECTIONS

Figure 38 shows the input and projections of the thalamus. The labels for the thalamic nuclei are paired with the areas to which they project and from which they receive input. The arrows on the lines to the labels reflect the direction of the impulses being considered.

Figure 39 represents a lateral and sagittal view of the cerebral cortex and shows schematically the areas of the cortex to which the various thalamic nuclei project. The labels indicate the name of the cortical area and the specific nuclei from which the projections come. Study of this diagram will enable you to get a better picture of the thalamic projection areas.

The numbers associated with the cortical areas in both figures refer to Brodmann's areas (see Sect. 63). The use of Brodmann's areas permits a more precise description of the specific parts of the cortex involved.

The areas of the central nervous system that project to the various thalamic nuclei are not shown in either figure; however, a reference is given for each structure mentioned so that you can easily find it on one of the drawings and thus become oriented.

All sensory input to the cortex is relayed through the thalamic nuclei. Not all thalamic nuclei project to the cortex, however. The **centromedian nucleus** (6–38), for example, receives fibers from the reticular nuclei of the brain stem, but the evidence for projections from the centromedian nucleus to the cortex is, at best, equivocal. It is also true that most of the cortical areas that receive fibers from the various nuclei of the thalamus also send fibers back to those nuclei. Thus, all of the thalamocortical tracts are two-way bundles.

The **anterior nuclei** (11–38) receive fibers from the **mammillothalamic tract** (12–38) and from the **fornix** (13–38). They project to most of the **cingulate gyrus** (41–38), (16–39). Functionally, these anterior nuclei are probably related to olfaction.

The **dorsomedial nucleus** (1–38), (15–39) can be functionally divided into two portions. The more medial part receives fibers from the **amygdala** (2–38) and projects fibers to the **hypothalamus** (3–38) and the **lateral preoptic region** (4–38). It does not send fibers to the cortex. The more lateral section of the dorsomedial nucleus receives fibers from other thalamic nuclei and sends projections to a large portion of the **prefrontal cortex** (5–38), (14–39). This nucleus is an important relay station between the hypothalamus and the prefrontal cortex and is involved in visceral impulses.

The **centromedian nucleus** (6–38) and the other intralaminar nuclei appear to receive fibers from the **globus pallidus** (8–38) and, as indicated earlier, the reticular formation. The centromedian nucleus also receives fibers from the premotor cortex, Brodmann's areas 6 and 8. Although the exact projections are in doubt, there appears to be good evidence that it projects to the **putamen** (10–38). It does not have direct cortical projections.

The **lateral dorsal nucleus** (40–38), (18–39) sends fibers to the **cingulate gyrus** (16–39). The origin of the input fibers to this nucleus has not yet been determined.

The **lateral posterior nucleus** (15–38), (6–39) receives connections from the other thalamic nuclei and sends fibers to the parietal lobe, specifically to the **superior parietal lobule** (17–38), (5–39).

The **pulvinar** (18–38), (11–39) receives input from other nuclei in the thalamus, particularly from the **lateral geniculate body** (29–38), (20–39) and the **medial geniculate body** (24–38), (13–39). The cortical projections from the pulvinar include the **inferior parietal lobule** (21–38), (10–39), the **posterior part of the temporal lobe** (22–38), (10–39), and the visual association areas of the occipital cortex, Brodmann's areas 18 and 19.

The **ventral anterior nucleus** (46–38), (2–39) projects to the **premotor area** (45–38), (1–39), Brodmann's area 6, although some anatomists dispute this finding. It also sends fibers to the **corpus striatum** (48–38) (see also Sect. 36) and receives fibers from the **globus pallidus** (49–38) and the **substantia nigra** (50–38).

The **ventral lateral nucleus** (42–38), (4–39) is a relay station between the cerebellum and the cortex. It receives fibers from the **superior cerebellar peduncle** (43–38) and sends them on to the **precentral gyrus** (3–39) of the cortex, with a few going to the **premotor area** (47–38), (1–39).

The **ventral posterior nucleus** (33,37–38), (8,9–39) is the terminal point for the principal sensory pathways. The **spinothalamic tracts** (34–38) and the **medial lemniscus** (35–38) end in this nucleus. The ventral posterior nucleus projects to the somesthetic cortex in the **postcentral gyrus** (39–38), (7–39).

The **ventral posterior medial nucleus** (37–38), (9–39) also receives sensory fibers, particularly from the **secondary trigeminal pathways** (38–38). It also projects to the somesthetic cortex.

The **lateral geniculate body** (29–38), (20–39), which is associated with vision, receives fibers from the **optic**

(Continued)

Figure 38

Schematic of the nuclei of the thalamus showing the principal input and areas of projection.

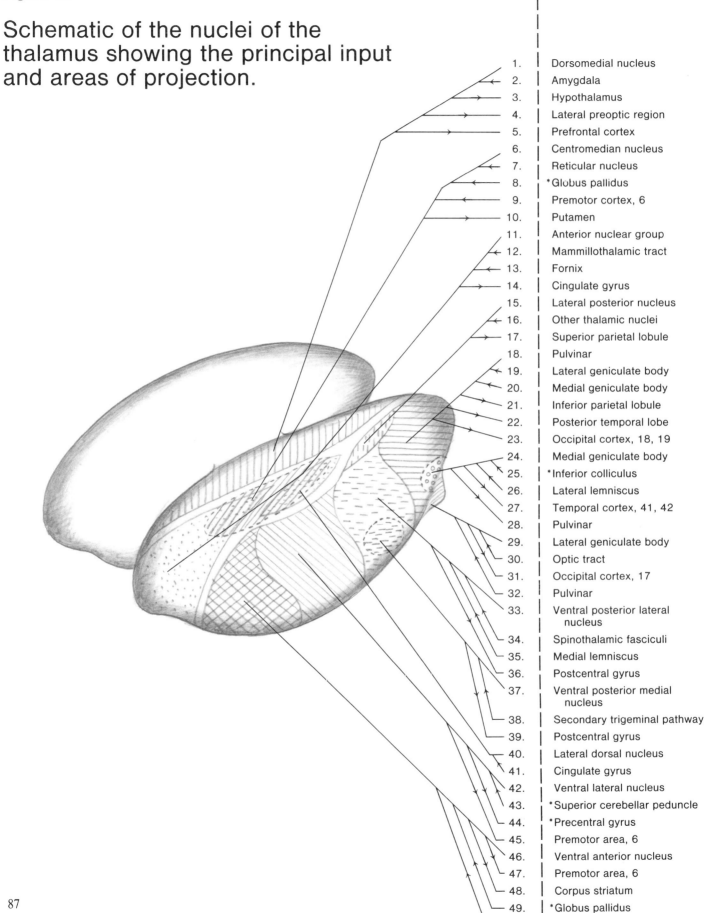

1. Dorsomedial nucleus
2. Amygdala
3. Hypothalamus
4. Lateral preoptic region
5. Prefrontal cortex
6. Centromedian nucleus
7. Reticular nucleus
8. *Globus pallidus
9. Premotor cortex, 6
10. Putamen
11. Anterior nuclear group
12. Mammillothalamic tract
13. Fornix
14. Cingulate gyrus
15. Lateral posterior nucleus
16. Other thalamic nuclei
17. Superior parietal lobule
18. Pulvinar
19. Lateral geniculate body
20. Medial geniculate body
21. Inferior parietal lobule
22. Posterior temporal lobe
23. Occipital cortex, 18, 19
24. Medial geniculate body
25. *Inferior colliculus
26. Lateral lemniscus
27. Temporal cortex, 41, 42
28. Pulvinar
29. Lateral geniculate body
30. Optic tract
31. Occipital cortex, 17
32. Pulvinar
33. Ventral posterior lateral nucleus
34. Spinothalamic fasciculi
35. Medial lemniscus
36. Postcentral gyrus
37. Ventral posterior medial nucleus
38. Secondary trigeminal pathway
39. Postcentral gyrus
40. Lateral dorsal nucleus
41. Cingulate gyrus
42. Ventral lateral nucleus
43. *Superior cerebellar peduncle
44. *Precentral gyrus
45. Premotor area, 6
46. Ventral anterior nucleus
47. Premotor area, 6
48. Corpus striatum
49. *Globus pallidus
50. Substantia nigra

tract (30–38) and relays them to the **primary visual area** (19–39), Brodmann's area 17, in the occipital cortex. The **medial geniculate body** (20–38), (13–39), which is associated with hearing, receives fibers from the **inferior colliculus** (25–38) and the **lateral lemniscus** (26–38) and projects them to the auditory area of the temporal cortex, Brodmann's areas 41 and 42 (12–39).

Figure 39

Lateral and sagittal aspects of the cerebral cortex showing the areas to which the nuclei of the thalamus project.

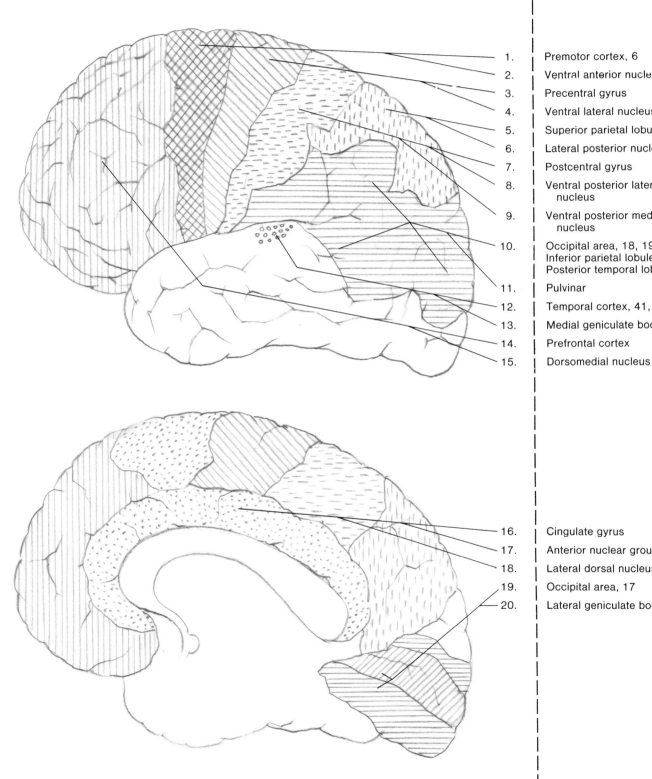

1. Premotor cortex, 6
2. Ventral anterior nucleus
3. Precentral gyrus
4. Ventral lateral nucleus
5. Superior parietal lobule
6. Lateral posterior nucleus
7. Postcentral gyrus
8. Ventral posterior lateral nucleus
9. Ventral posterior medial nucleus
10. Occipital area, 18, 19
    Inferior parietal lobule
    Posterior temporal lobe
11. Pulvinar
12. Temporal cortex, 41, 42
13. Medial geniculate body
14. Prefrontal cortex
15. Dorsomedial nucleus

16. Cingulate gyrus
17. Anterior nuclear group
18. Lateral dorsal nucleus
19. Occipital area, 17
20. Lateral geniculate body

THE BASAL GANGLIA

The basal ganglia are relatively large masses of gray matter located in the telencephalon. They are separated from the diencephalon by the internal capsule. The nomenclature of the basal ganglia tends to be confusing because the ganglia are grouped together in different ways. It is also difficult to get a good three-dimensional picture of these ganglia and their relationships to one another and to other parts of the brain because of their complex shape. The list of basal ganglia and Figures 40, 41, and 42 should help resolve both of these problems. You will get further practice in identifying the basal ganglia in the coronal (Figs. 43–46) and horizontal (Figs. 47–50) sections of the brain. The schematics in the following Figures 40, 41, and 42, however, should help to provide an immediate overall picture.

Although anatomists differ on the structures in the basal ganglia, the most commonly included structures and the manner in which they are most often grouped together are as follows:

        basal ganglia
            corpus striatum
                caudate nucleus
                internal capsule
                lenticular nucleus
                    putamen
                    globus pallidus
            amygdala
            claustrum

The **corpus striatum** (48–38), a general term for the basal ganglia exclusive of the **amygdala** (6–40) and the **claustrum** (6–41), receives its name from its striated appearance in histological sections. Bands of gray substance travel through the internal capsule connecting the **lenticular nucleus** (5–40) and the **caudate nucleus** (1–40), (1–41), (5–42).

The caudate nucleus has a relatively large anterior portion, known as the head, which is directly connected with the **putamen** (11–41), (11–42). The rest of the caudate nucleus forms an arch of continuously diminishing bulk curving around the thalamus and just lateral to it. The most slender part of the tail of the caudate proceeds in a rostral direction and terminates in the amygdala. Because the caudate nucleus forms an extensive arch, any coronal section of the middle of the brain will show two portions of it, the more dorsal body and the more ventral tail.

The lenticular nucleus, also referred to as the *lentiform nucleus* because it is shaped like a double convex lens, is composed of the putamen and the **globus pallidus** (13–41), (15–42). It is divided into sections by an **external medullary stria** (12–41), (13–42) and an **internal medullary stria** (15–41), (14–42). The external medullary stria separates the putamen from the globus pallidus, and the internal medullary stria divides the globus pallidus into an internal and an external section.

The putamen is bounded medially by the head of the caudate nucleus and the globus pallidus. Laterally it is separated from the claustrum by the **external capsule** (7–41), (10–42). The cytological structure of the caudate nucleus and the putamen is essentially the same. Both are composed of relatively small cells with a few intermediate-sized cells interspersed among them.

The globus pallidus, however, has quite a different cell structure. The cells tend to be large, multipolar neurons of the motor type. Also, large numbers of myelinated fibers run through the globus pallidus. The anterior and medial borders of the globus pallidus are formed by the **internal capsule** (4, 8,17–41), (7–42), which in horizontal section is shaped like a "V" with the point directed medially. The globus pallidus is separated from the putamen by the external medullary stria and is phylogenetically older than the rest of the corpus striatum. The globus pallidus is frequently referred to as the *pallidum*.

The functions of the claustrum are not known. It forms a relatively thin sheet of gray matter between the putamen and the **insula** (12–42). It is separated from the putamen by the external capsule and from the insula by the **extreme capsule** (16–43).

The amygdala, a complex mass of gray matter composed of as many as eight identifiable nuclei, lies in the dorsomedial portion of the temporal lobe. It has connections with the caudate nucleus and sends a large number of fibers to the hypothalamus via the **stria terminalis** (2–34c). Its functions are complex, and a number of its nuclei are involved in emotional responding.

The internal capsule is a broad band of fibers between the lenticular nucleus and the caudate nucleus. The portion of the internal capsule that separates the caudate nucleus from the lenticular nucleus is called the **anterior limb** (4–41), and that portion separating the thalamus from the lenticular nucleus is the **posterior limb** (17–41). The bend in the middle is called the **genu** (8–41).

(Continued)

Figure 40

Three-dimensional view of the basal ganglia.

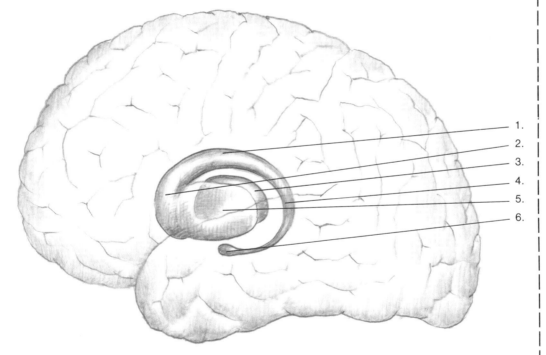

1.   Caudate nucleus
2.   Head of caudate nucleus
3.   *Thalamus
4.   Tail of caudate nucleus
5.   *Lenticular nucleus
6.   Amygdala

The anterior limb of the internal capsule contains fibers that connect the frontal lobe of the cortex with the thalamus and are called the **anterior thalamic radiation** (3–41). It also contains fibers from the motor area of the cortex to the cerebral peduncle and the pontine nuclei that are called the **frontopontile tract** (2–41). **Corticobulbar tracts** (9–41) run through the genu. The anterior portion of the posterior limb of the internal capsule contains motor fibers that descend to the **cerebral peduncle** (6–7). They then go to the **pyramid** (22–20) of the medulla oblongata and of the cord where they are called the **corticospinal tract** (10, 16–41). Some of the fibers connect the cortex with the red nucleus and are called the **corticorubral tract** (14–41). The caudal portion of the posterior limb contains the **superior thalamic radiation** (22–41), the major fiber bundle from the thalamus to the somesthetic cortex. The most posterior aspect of the posterior portion of the internal capsule contains the **auditory radiation** (18–41), which connects the **medial geniculate body** (24–38) with the temporal cortex, and the **optic radiation** (20–41), which connects the **lateral geniculate body** (29–38) with the **calcarine fissure** (25–3) in the occipital lobe.

Figure 41

Horizontal section through the basal ganglia.

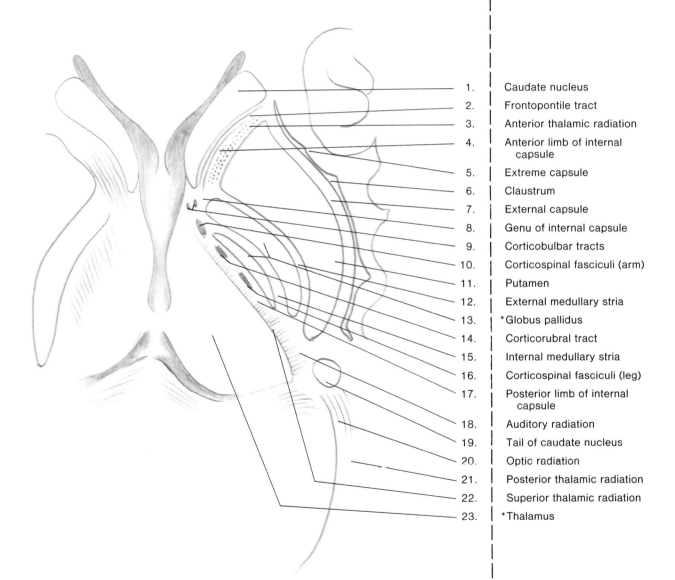

1. Caudate nucleus
2. Frontopontile tract
3. Anterior thalamic radiation
4. Anterior limb of internal capsule
5. Extreme capsule
6. Claustrum
7. External capsule
8. Genu of internal capsule
9. Corticobulbar tracts
10. Corticospinal fasciculi (arm)
11. Putamen
12. External medullary stria
13. *Globus pallidus
14. Corticorubral tract
15. Internal medullary stria
16. Corticospinal fasciculi (leg)
17. Posterior limb of internal capsule
18. Auditory radiation
19. Tail of caudate nucleus
20. Optic radiation
21. Posterior thalamic radiation
22. Superior thalamic radiation
23. *Thalamus

Figure 42

Coronal section through the basal ganglia including the amygdala.

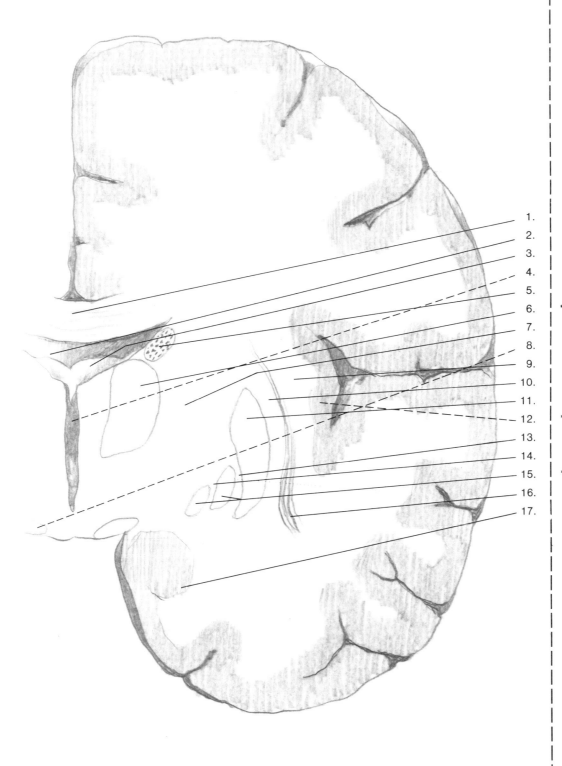

1. Corpus callosum
2. Lateral ventricle
3. Fornix
4. Third ventricle
5. Caudate nucleus
6. *Thalamus
7. Internal capsule
8. Optic tract
9. Extreme capsule
10. External capsule
11. Putamen
12. *Insula
13. External medullary stria
14. Internal medullary stria
15. *Globus pallidus
16. Claustrum
17. Amygdala

HEMISPHERE SECTIONS

Figures 43–50 are sections through the central portions of the hemispheres. The first four are coronal sections, and the last four are horizontal sections. They should be studied individually and in sequence. That is, you should learn the relationships between the various structures in each section and also follow a given structure through all of the sections in order to note the changes in the shapes of the individual structures and the manner in which the relationships between them change. This method of study will contribute further to a good three-dimensional conception of the components of the brain.

The individual nuclei in the thalamus and the hypothalamus have not been labeled in these sections because they have been presented in detail elsewhere. Also, many of the items labeled in this series of drawings are not discussed in this portion of the text because they have been more fully covered in other sections. The Indexes will provide ready access to those portions of the text.

In order to avoid undue complexity in the system of labels, most of the sulci and gyri have not been included on the labels. Enough have been covered, however, so that orientation is possible. The **cingulate gyrus** (1–43), (1–44), (1–45), (1–46), (2–47), (2–48), (2–49) and the **corpus callosum** (2–43), (2–44), (34–45), (2–46) for example, are labeled in each section as they occur, as are the **lateral fissure** (15–43), (15–44), (21–45), (17–46), (4–47), (18–48), (12–49), (10–50), the **limiting fissure** (13–43), (12–44), (13–45), (12–46), (8–48), (6–49), (6–50), and the **insula** (17–43), (13–44), (15–45), (15–46), (15–48), (11–49), (9–50). In the case of the last two items, these sections provide a three-dimensional view that is difficult to obtain in other types of representations. The structures at the base of the brain—the **interpeduncular fossa** (36–46), or *posterior perforated substance,* the **infundibulum** (33–44), the **tuber cinereum** (32–44), the **pons** (37–46), and the **cerebellum** (39–49), (35–50)—which have been shown in earlier figures, will facilitate orientation when using these sections.

Figure 43 represents a coronal section through the **anterior commissure** (20–43), (15–50), a large bundle of white fibers that crosses the midline and establishes connections between portions of the two temporal lobes. It also links the olfactory bulb on one side with the olfactory tract on the opposite side.

The basal ganglia shown in three dimensions in Figure 40 can be readily traced through these sections. The **globus pallidus** (19–43), (22–44), (22–45), (21–46), (20–49), (19–50) can easily be followed through each of the sections; it will be noted that it shifts to a more lateral position as it decreases in size in the consecutive coronal sections. The **external medullary stria** (12–43), (19–44), (20–45), (19–46), (15–49), (14–50), which separates the globus pallidus from the putamen, is evident in the first section. However, the **internal medullary stria** (21–44), (23–45), (19–49), (18–50), which divides the globus pallidus into two portions, does not become apparent until the second section. The **substantia innominata** (28–45) is shown just below the **lenticular nucleus.**

The relationship between the **putamen** (11–43), (18–44), (16–45), (14–46), (14–48), (13–49), (8–50), and the **head of the caudate nucleus** (7–43), (8–44), (7–48), (9–49), (5–50) can also be clearly seen by following the figures. In the more anterior section (Fig. 43) they are separated, but they have connections in the second section (Fig. 44) that are lost as the head of the caudate thins out and curves dorsally to form the body. The connection between these two nuclei is also shown in the horizontal section in Figure 50.

The **tail of the caudate nucleus** (27–45), (26–46), (23–48), (32–49), (24–50) can be seen to lie posterior to the **amygdala** (22–43), (24–44), as shown in Figure 40. The successive sections also provide a better idea of the shape and course of the tail of the caudate.

The basic shape of the **thalamus** (9–45), (8–46), (21–48), (26–49), (17–50) can also be deduced by noting its position in each section. The individual nuclei have not been labeled because they are covered in Figure 36. However, the **internal medullary lamina** (9–46) is shown, as is the **external medullary lamina** (13–46). The **intermediate mass** (8–45), (25–49) is indicated in one coronal and one horizontal section, and the **pulvinar** (30–49) is labeled in Figure 36. The **zona incerta** (12–45), (23–46) is shown just below the thalamus.

This series of figures will also clarify the relationships among the **claustrum** (14–43), (20–44), (18–45), (22–46), (13–48), (21–49), (13–50), the other masses of gray matter, and the large fiber tracts that course among them. The three sections of the **internal capsule**—the **anterior limb** (9–43), (10–48), the **genu** (14–44), and the **posterior limb** (10–45), (16–46), (20–48), (24–49)—are shown in both coronal and horizontal sections. The same is true for the **external capsule** (18–43), (16–44), (14–45), (18–46), (16–48), (18–49), and the **extreme capsule** (16–43), (17–44), (24–45), (20–46), (12–48), (17–49). The

(Continued)

Figure 43

Coronal section of the cerebral hemisphere through the anterior commissure.

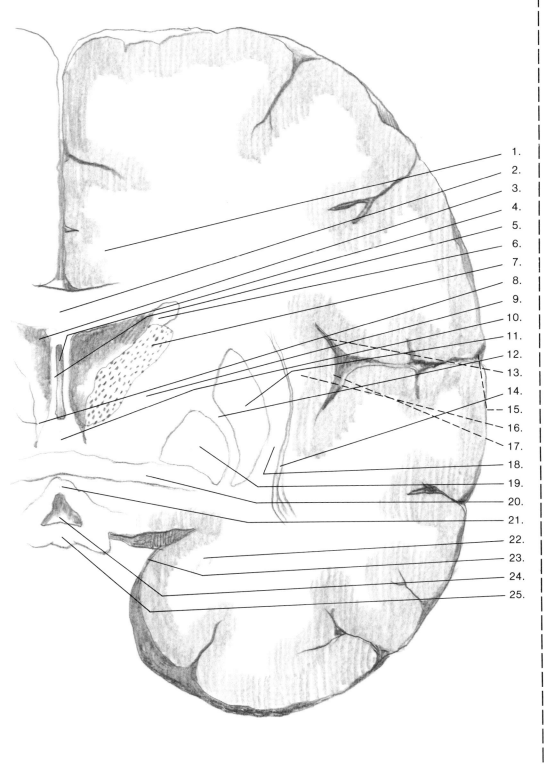

1. Cingulate gyrus
2. Corpus callosum, body
3. Lateral ventricle, anterior horn
4. Cavity of septum pellucidum
5. Septum pellucidum
6. Superior occipitofrontal bundle
7. Caudate nucleus, head
8. Stria terminalis
9. Internal capsule, anterior limb
10. Precommissural fornix
11. Putamen
12. External medullary stria
13. *Limiting fissure
14. Claustrum
15. *Lateral fissure
16. Extreme capsule
17. *Insula
18. External capsule
19. *Globus pallidus
20. Anterior commissure
21. Preoptic region
22. Amygdala
23. Uncus
24. Third ventricle
25. Optic chiasm

various aspects of the great interhemisphere band of fibers are also shown, including the **body of the corpus callosum** (2–43), (2–44), (34–45), (2–46), the **genu** (3–48), (3–49), (1–50), the **rostrum** (4–50), and the **splenium** (27–48). The horizontal section above the corpus callosum shows the white matter in an oval shape. This was called the **centrum semiovale** (3–47) by the early anatomists.

In addition to the structures already covered, many of the other components of the extrapyramidal system (see Sect. 58) can be traced through these hemisphere sections, including the **subthalamic nuclei** (19–45), (24–46), the **substantia nigra** (32–46) (22–50), the **cerebral peduncle** (26–45), (33–46), (27–50), and the **red nucleus** (31–46), (29–50).

A three-dimensional view of the relationships among the **mammillary bodies** (25–45), the **fornix** (10–43), (9–44), (6–45), (6–46), (22–48), (16–49), (16–50), and the **hippocampus** (26–44), (32–45), (27–46), (26–48), (35–49), (28–50) is given in Figure 54, and these structures are discussed in greater detail in Section 40. It is helpful, however, to trace these structures through the sections and learn their relationships to other structures. Use Figure 54 to get the overall picture, and then return to this portion of the book to locate each structure in each section. Note particularly the **fimbria** (28–46), (33–49), (26–50), the **fusiform gyrus** (31–44), (34–46), and the **uncus** (23–43), (28–44), (33–45), (35–46).

The same general procedure will be useful in the study of the relation of the lateral ventricles. The three-dimensional schematic in Figure 61 will provide an overall picture, and then the individual parts of the ventricles can be identified in relationship to the rest of the brain. The following portions are shown in these sections: **lateral ventricle, anterior horn** (3–43), (5–48), (5–49); **lateral ventricle, body** (5–44), (2–45), (3–46); **lateral ventricle, inferior horn** (25–44), (31–45), (25–46); and **lateral ventricle, posterior horn** (30–48). The **third ventricle** (24–43), (29–44), (11–46), (28–49), (2–50) is also shown, as are the **cerebral aqueduct** (32–50) and the **interventricular foramen** (22–49).

The **choroid plexus of the lateral ventricle** (5–46), (24–48), (25–50) and the **choroid plexus of the third ventricle** (10–46) produce the greater part of the cerebrospinal fluid (see Sect. 44). The relationship between the choroid plexuses and the ventricles and the other brain structures can be appreciated by the study of this series of sections.

The **septum pellucidum** (5–43), (4–44), (9–48), (8–49) consists of two thin verticle lamina that separate the two lateral ventricles. These lamina are separated from each other by the **cavity of the septum pellucidum** (4–43), (3–44), (6–48). The septum pellucidum fills the triangular interval between the corpus callosum and the fornix.

Several important fiber bundles appear in almost every section, and the relative three-dimensional positions of each can thus be readily determined. The **superior occipitofrontal bundle** (6–43), (7–44), (3–45), (4–48), (4–49), (3–50) is one of the long association bundles of the hemisphere that connects the occipital and temporal regions with the frontal lobe and the insula. The **stria terminalis** (8–43), (6–44), (5–45), (29–46), (25–48), (31–49), (23–50) runs from the amygdaloid nucleus along the entire medial border of the caudate nucleus to terminate in the anterior hypothalamic region. The **stria medullaris** (7–45), (27–49) courses from the preoptic region of the hypothalamus to the **habenula** (29–49). The **mammillothalamic tract** (10–44), (11–45), (23–49) arises in the **mammillary bodies** (25–45) and terminates in the anterior thalamic nuclei.

A number of components of the visual system that are considered in detail elsewhere should be noted. These include the **superior colliculus** (37–49), the **lateral geniculate body** (31–50), the **visual radiations** (29–48), (38–49), (30–50), the **anterior calcarine fissure** (28–48), and the **posterior calcarine fissure** (31–48). At the base of the brain the sections go through the **optic nerve** [II] (3–66), the **optic chiasm** (25–43), and the **optic tract** (30–44), (29–45), (30–46), (21–50). The auditory components include the **inferior colliculus** (34–50), the **medial geniculate body** (33–50), and the **auditory radiation** (20–50).

Figure 44

Coronal section of the cerebral hemisphere through the tuber cinereum.

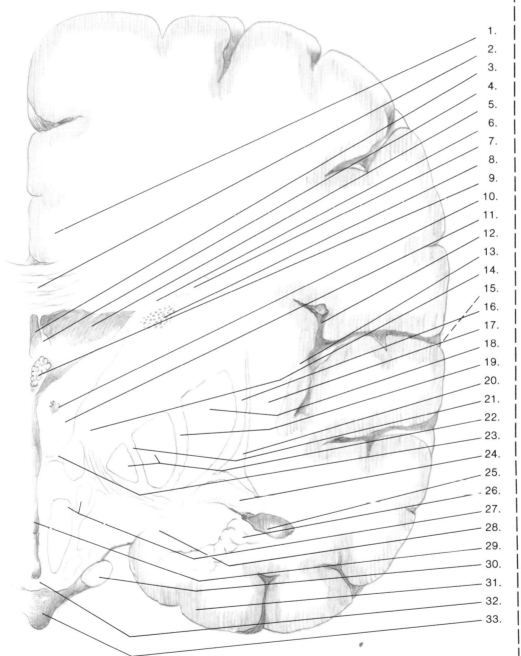

1.   Cingulate gyrus
2.   Corpus callosum
3.   Cavity of septum pellucidum
4.   Septum pellucidum
5.   Lateral ventricle, body
6.   Stria terminalis
7.   Superior occipitofrontal bundle
8.   Caudate nucleus, head
9.   Fornix
10. Mammillothalamic tract
11. *Thalamus
12. *Limiting fissure
13. *Insula
14. Internal capsule, genu
15. *Lateral fissure
16. External capsule
17. Extreme capsule
18. Putamen
19. External medullary stria
20. Claustrum
21. Internal medullary stria
22. *Globus pallidus
23. Hypothalamic sulcus
24. Amygdala
25. Lateral ventricle, inferior horn
26. *Hippocampus
27. Hypothalamus
28. Uncus
29. Third ventricle
30. Optic tract
31. *Fusiform gyrus
32. Tuber cinereum
33. Infundibulum

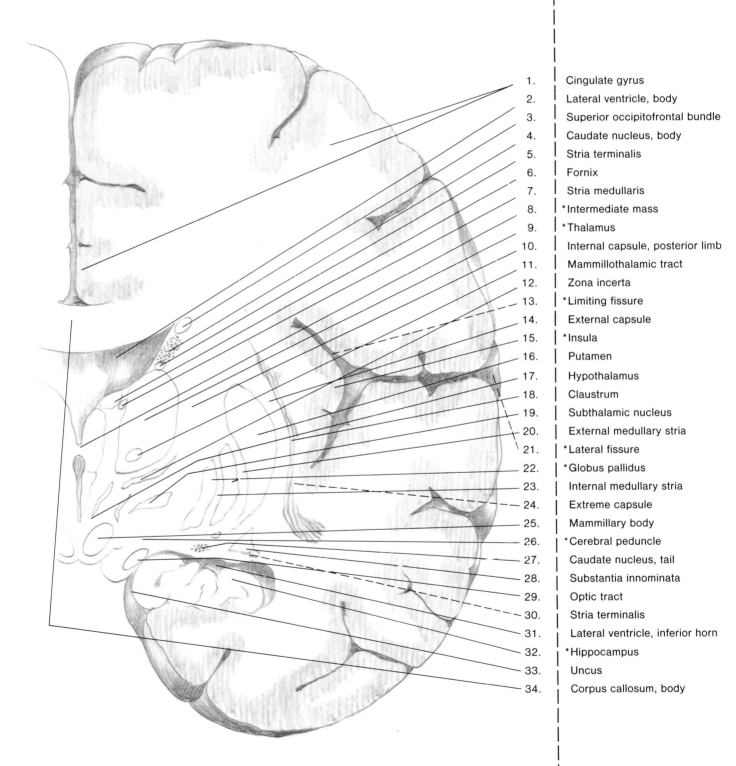

Figure 45

Coronal section of the cerebral hemisphere through the mammillary bodies.

1. Cingulate gyrus
2. Lateral ventricle, body
3. Superior occipitofrontal bundle
4. Caudate nucleus, body
5. Stria terminalis
6. Fornix
7. Stria medullaris
8. *Intermediate mass
9. *Thalamus
10. Internal capsule, posterior limb
11. Mammillothalamic tract
12. Zona incerta
13. *Limiting fissure
14. External capsule
15. *Insula
16. Putamen
17. Hypothalamus
18. Claustrum
19. Subthalamic nucleus
20. External medullary stria
21. *Lateral fissure
22. *Globus pallidus
23. Internal medullary stria
24. Extreme capsule
25. Mammillary body
26. *Cerebral peduncle
27. Caudate nucleus, tail
28. Substantia innominata
29. Optic tract
30. Stria terminalis
31. Lateral ventricle, inferior horn
32. *Hippocampus
33. Uncus
34. Corpus callosum, body

Figure 46

Coronal section of the cerebral hemisphere through the red nucleus.

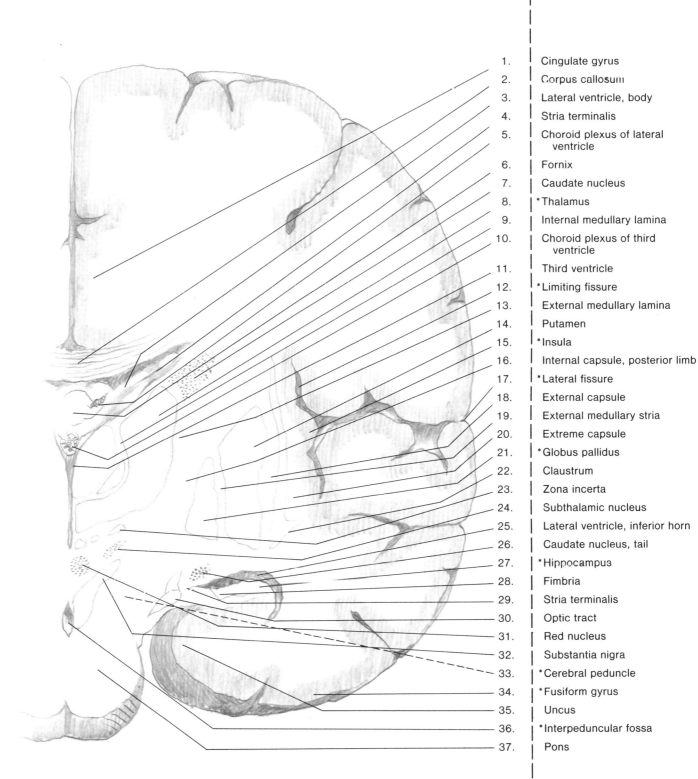

1. Cingulate gyrus
2. Corpus callosum
3. Lateral ventricle, body
4. Stria terminalis
5. Choroid plexus of lateral ventricle
6. Fornix
7. Caudate nucleus
8. *Thalamus
9. Internal medullary lamina
10. Choroid plexus of third ventricle
11. Third ventricle
12. *Limiting fissure
13. External medullary lamina
14. Putamen
15. *Insula
16. Internal capsule, posterior limb
17. *Lateral fissure
18. External capsule
19. External medullary stria
20. Extreme capsule
21. *Globus pallidus
22. Claustrum
23. Zona incerta
24. Subthalamic nucleus
25. Lateral ventricle, inferior horn
26. Caudate nucleus, tail
27. *Hippocampus
28. Fimbria
29. Stria terminalis
30. Optic tract
31. Red nucleus
32. Substantia nigra
33. *Cerebral peduncle
34. *Fusiform gyrus
35. Uncus
36. *Interpeduncular fossa
37. Pons

Figure 47

Horizontal section of the cerebral hemisphere through the centrum semiovale.

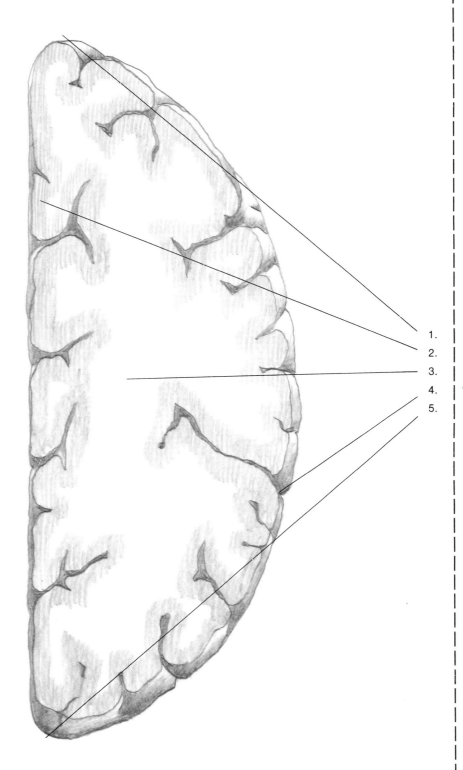

1. | Frontal pole
2. | Cingulate gyrus
3. | Centrum semiovale
4. | *Lateral fissure
5. | Occipital pole

Figure 48

Horizontal section of the cerebral hemisphere through the dorsal thalamus.

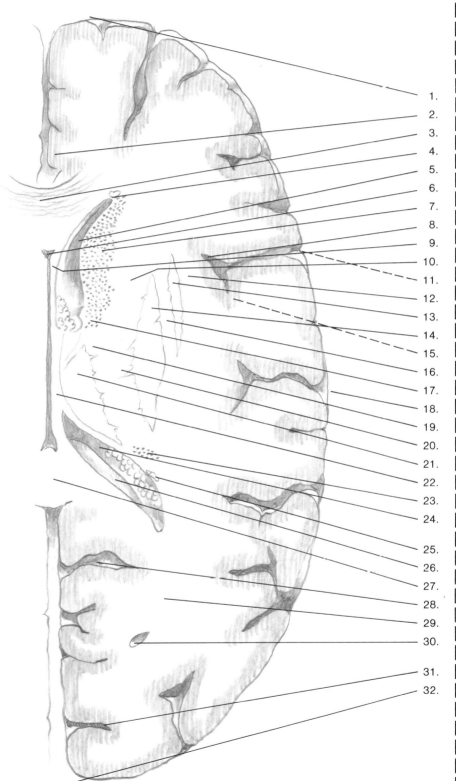

1.   Frontal pole
2.   Cingulate gyrus
3.   Corpus callosum, genu
4.   Superior occipitofrontal bundle
5.   Lateral ventricle, anterior horn
6.   Cavity of septum pellucidum
7.   Caudate nucleus, head
8.   *Limiting fissure
9.   Septum pellucidum
10.   Internal capsule, anterior limb
11.   *Central fissure
12.   Extreme capsule
13.   Claustrum
14.   Putamen
15.   *Insula
16.   External capsule
17.   Stria terminalis
18.   *Lateral fissure
19.   Internal capsule, genu
20.   Internal capsule, posterior limb
21.   *Thalamus
22.   Fornix, body
23.   Caudate nucleus, tail
24.   Choroid plexus of lateral ventricle
25.   Stria terminalis
26.   *Hippocampus
27.   Corpus callosum, splenium
28.   Anterior calcarine sulcus
29.   Visual radiations
30.   Lateral ventricle, posterior horn
31.   Posterior calcarine sulcus
32.   Occipital pole

Figure 49

Horizontal section of the cerebral hemisphere through the internal capsule.

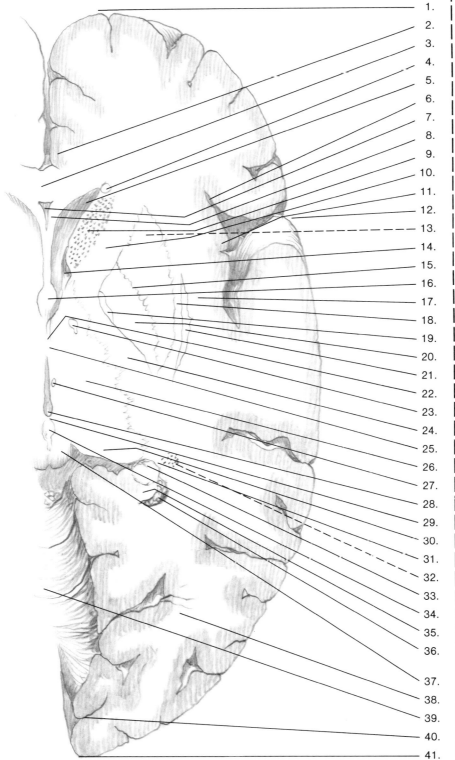

1. Frontal pole
2. Cingulate gyrus
3. Corpus callosum, genu
4. Superior occipitofrontal bundle
5. Lateral ventricle, anterior horn
6. *Limiting fissure
7. Cavity of septum pellucidum
8. Septum pellucidum
9. Caudate nucleus, head
10. Internal capsule, anterior limb
11. *Insula
12. *Lateral fissure
13. Putamen
14. Stria terminalis
15. External medullary stria
16. Fornix
17. Extreme capsule
18. External capsule
19. Internal medullary stria
20. *Globus pallidus
21. Claustrum
22. *Interventricular foramen
23. Mammillothalamic tract
24. Internal capsule, posterior limb
25. *Intermediate mass
26. *Thalamus
27. Stria medullaris
28. Third ventricle
29. Habenula
30. Pulvinar
31. Stria terminalis
32. Caudate nucleus, tail
33. Fimbria
34. *Pineal body
35. *Hippocampus
36. Lateral ventricle, posterior horn
37. *Superior colliculus
38. Visual radiations
39. Cerebellum
40. Posterior calcarine fissure
41. Occipital pole

Figure 50

Horizontal section of the cerebral hemisphere through the basal ganglia.

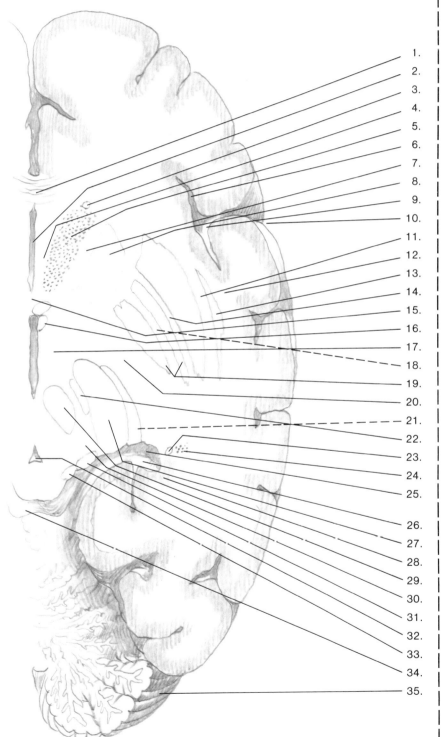

1.	Corpus callosum, genu
2.	Third ventricle
3.	Superior occipitofrontal bundle
4.	Rostrum of corpus callosum
5.	Caudate nucleus, head
6.	*Limiting fissure
7.	Internal capsule, anterior limb
8.	Putamen
9.	*Insula
10.	*Lateral fissure
11.	External capsule
12.	Extreme capsule
13.	Claustrum
14.	External medullary stria
15.	Anterior commissure
16.	Fornix
17.	*Thalamus
18.	Internal medullary stria
19.	*Globus pallidus
20.	Auditory radiations
21.	Optic tract
22.	Substantia nigra
23.	Stria terminalis
24.	Caudate nucleus, tail
25.	Choroid plexus of inferior horn of lateral ventricle
26.	Fimbria
27.	*Cerebral peduncle
28.	*Hippocampus
29.	Red nucleus
30.	Visual radiations
31.	Lateral geniculate body
32.	*Cerebral aqueduct
33.	Medial geniculate body
34.	*Inferior colliculus
35.	Cerebellar hemisphere

THE ASSOCIATION FIBERS
OF THE CEREBRAL HEMISPHERE

The medullary substance, or white matter, of the hemispheres consists of myelinated nerve fibers. It varies in thickness from the relatively massive **centrum semiovale** (3–47) to the thin band of white matter, the **extreme capsule** (16–43), (12–48), that separates the claustrum from the insula.

There are three basic kinds of fibers: projection fibers, which go to and from the hemispheres; association fibers, which connect one part of the hemisphere with another; and commissural fibers, which form connections between the two hemispheres. The projection fibers have been illustrated in earlier sections and will be covered in more detail later. They come from all portions of the cortex and enter the medullary substance to form the **corona radiata** (1–8). The fibers then converge on the **internal capsule** (7–42), (2–9) and descend into the brain stem. They ultimately form the tracts, or fasciculi, of the spinal cord. They also course through the **external capsule** (10–42), (16–48). The association fibers are illustrated in Figure 51, and the commissural fibers are diagramed in Section 39.

The association fibers are of two kinds. The short association fibers are *subcortical fibers* and connect adjacent gyri. Since they curve under the sulci, they form half circles and are thus called **arcuate fibers** (4–51). The *intracortical fibers* are the long association fibers, which run deep in the hemisphere in bundles of considerable size.

The **cingulum fasciculus** (5–51) is a significant bundle of fibers contained within the cingulate gyrus. It arches around the corpus callosum, beginning ventral to the rostrum of the corpus callosum and curving up and around the genu. It proceeds over the body of the corpus callosum, curving around the splenium to terminate and establish connections in the hippocampal gyrus.

The **uncinate fasciculus** (3–51) establishes connections between the gyri of the frontal lobe and the anterior portion of the temporal lobe. It curves around the stem of the lateral fissure of the hemisphere.

The **superior longitudinal fasciculus** (1–51), or *fasciculus arcuatus,* connects many portions of the hemisphere. It originates in the frontal lobe and establishes connections between that area and the parietal, occipital, and temporal areas of the cortex. It contains many short fibers.

The **inferior longitudinal fasciculus** (7–51) connects the temporal and the occipital lobes. It runs along the lateral walls of the inferior and posterior horns of the lateral ventricle. It also establishes some connections with the parietal lobe.

The **inferior occipitofrontal fasciculus** (2–51) is located in the inferior portion of the extreme capsule and runs from the occipital to the frontal lobes.

The **perpendicular fasciculus** (6–51) makes connections between the **fusiform gyrus** (19–6) and the **inferior parietal lobule** (15–3). It runs vertically in the anterior portion of the occipital lobe and is sometimes called the *vertical occipital fasciculus.*

The **extreme capsule** (16–43), (12–48) is considered by some anatomists to be composed of association fibers that connect parts of the insular cortex as well as make connections between the insula and the claustrum.

Figure 51

Principal association fasciculi of the cerebral hemisphere.

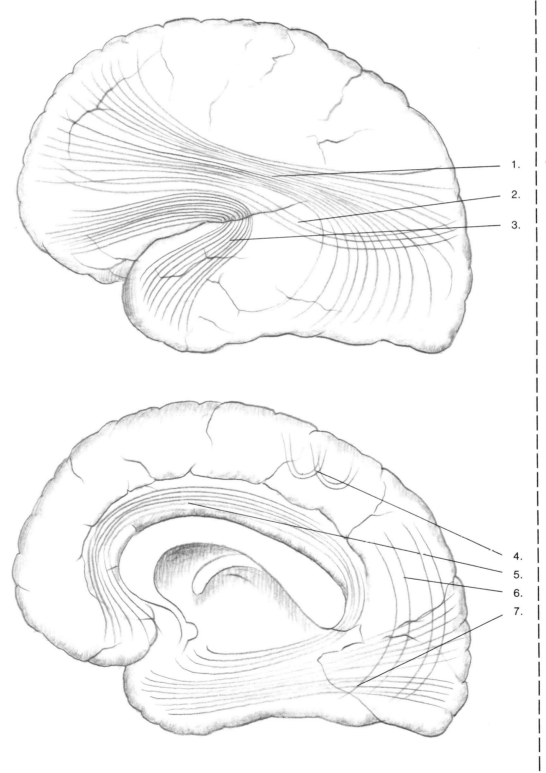

1. *Superior longitudinal fasciculus
2. Inferior occipitofrontal fasciculus
3. Uncinate fasciculus

4. Arcuate fibers, cerebral
5. Cingulum fasciculus
6. *Perpendicular fasciculus
7. Inferior longitudinal fasciculus

THE COMMISSURAL FIBERS OF THE CEREBRAL HEMISPHERES

Figure 52 shows the major commissural fibers, which connect the two hemispheres. The principal association-fiber bundles that run through this portion of the hemisphere are also shown in order to enhance the general picture of these connections. Those fasciculi have been discussed in Section 38.

The most important connection between the two hemispheres is, of course, the **corpus callosum** (6–52). It is composed primarily of medullated fibers, but some of its fibers are unmedullated. Most of the fibers are commissural, connecting the two halves of the brain. However, the corpus callosum also contains collaterals of the association fasciculi and the projection fibers. The radiations of the corpus callosum make up most of the bulk of the **centrum semiovale** (3–47). Relatively few of the fibers of the corpus callosum are located in the visual area near the calcarine fissures.

This large band of fibers constitutes the roof of the lateral ventricles. The general shape of the corpus callosum can best be seen in the midsagittal section of the hemisphere (15–5), (6–52). It will be remembered that the most anterior end is the **genu** (2–5), which curves under in a ventral direction to form the **rostrum** (21–5). The posterior portion is the **splenium** (16–5). The anterior radiations of the corpus callosum, which curve from the genu into the frontal lobes, are called the *anterior forceps*. Those radiations that curve back into the occipital lobe are the *posterior forceps*. The body of the corpus callosum between the anterior and posterior forceps is referred to as the *tapetum*.

The **anterior commissure** (13–52) has already been described in Section 37. The **posterior commissure** (12–29) probably connects the nuclei of the reticular substance of the midbrain. It is a rounded band of white fibers that crosses the midline just anterior to the tectum.

The **hippocampal commissure** (4–54) is pictured in detail in Figure 54 and is discussed in the section on the hippocampus. Its fibers course through the **fimbria** (15–53), passing below the splenium of the corpus callosum.

Figure 52

Principal commissural fibers of the cerebral hemispheres.

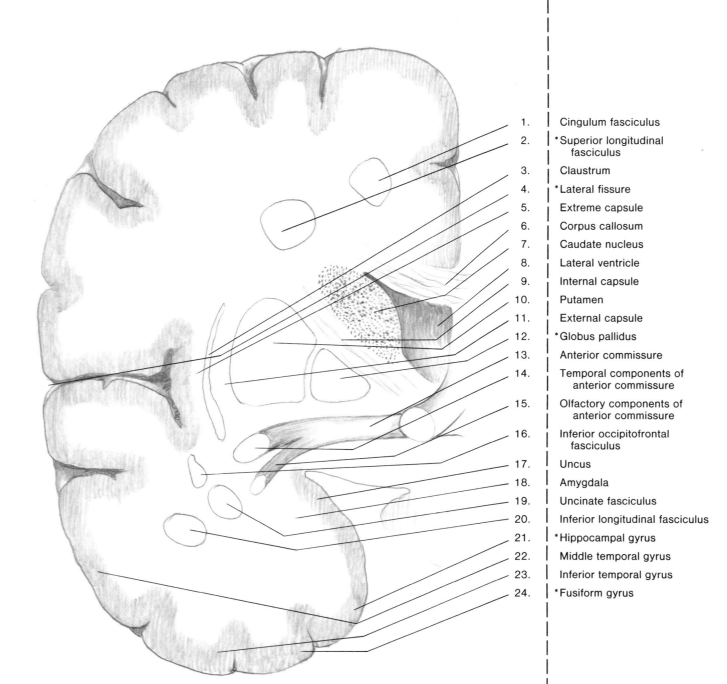

1.   Cingulum fasciculus
2.   *Superior longitudinal fasciculus
3.   Claustrum
4.   *Lateral fissure
5.   Extreme capsule
6.   Corpus callosum
7.   Caudate nucleus
8.   Lateral ventricle
9.   Internal capsule
10.   Putamen
11.   External capsule
12.   *Globus pallidus
13.   Anterior commissure
14.   Temporal components of anterior commissure
15.   Olfactory components of anterior commissure
16.   Inferior occipitofrontal fasciculus
17.   Uncus
18.   Amygdala
19.   Uncinate fasciculus
20.   Inferior longitudinal fasciculus
21.   *Hippocampal gyrus
22.   Middle temporal gyrus
23.   Inferior temporal gyrus
24.   *Fusiform gyrus

THE RHINENCEPHALON

The next four figures are concerned with the rhinencephalon and its component parts. Figure 53 is a schematic of the entire rhinencephalon with some of the more important interconnections included. The next two figures show the hippocampus, which is difficult to visualize. In Figure 55 it is shown in situ with some of the overlying material dissected away. In Figure 54 the hippocampus and its associated structures are shown in three dimensions. Finally, in Figure 56 a cross section of the hippocampus is shown. All four drawings will be discussed together. It will be useful for you to locate the structure being discussed in each of the drawings. Many of the structures associated with the rhinencephalon may be clearly seen on the ventral aspect of the cerebral hemispheres (Fig. 6). This drawing will also be referred to in the subsequent discussion.

The term *rhinencephalon* derives from Greek and means "nose brain" or "smell brain." It includes those portions of the cerebrum that are involved in the reception and transmission of olfactory stimuli. The relevant structures are the olfactory bulb, tract, and trigone and the pyriform area. The hippocampal formation and the fornix, as well as the subcallosal gyrus and supracallosal gyrus, are also included.

The olfactory pathways will be discussed in more detail in connection with the sensory systems (see Sect. 56). This section will include only a general description of the structures. The **olfactory bulb** (22–53) is flattened and ovoid in shape and rests on the cribriform plate of the ethmoid bone of the skull. It is reddish-gray in color. The bulb diminishes in size posteriorly and merges into the **olfactory tract** (18–53), which is triangular in shape on cross section. It lies in the olfactory sulcus. As the tract proceeds in a posterior direction it divides into the **lateral olfactory stria** (21–53) and the **medial olfactory stria** (10–53).

Between the two olfactory striae lies the **olfactory trigone** (12–53). The **anterior perforated substance** (20–53) is located just posterior to the olfactory trigone. It receives fibers from the two olfactory striae.

The *pyriform area* is composed of the anterior portion of the **hippocampal gyrus** (20–6), the **uncus** (14–53), (27–5), the **lateral olfactory stria** (21–53), and a thin layer of gray matter that covers it and that is known as the **olfactory gyrus.**

The *hippocampal formation* is a relatively complex structure that occupies a significant part of the central portion of each of the hemispheres. It includes the subcallosal and supracallosal gyri, the diagonal band of Broca, the longitudinal striae of the corpus callosum, the dentate gyrus, and the hippocampus itself.

The **subcallosal gyrus** (8–53), which is also called the *paraterminal body,* is a thin sheet of gray matter that curves around the **genu of the corpus callosum** (2–5). It is continuous with the **supracallosal gyrus** (2–53), which separates the corpus callosum and the cingulate gyrus. The supracallosal gyrus is also called the *indusium griseum.*

The lateral and medial longitudinal striae constitute the white substance of the vestigial supracallosal gyrus. They curve around the corpus callosum and in their anterior aspect enter the subcallosal gyrus. They emerge from that gyrus as the **diagonal band of Broca** (13–53), which forms the posterior border of the **anterior perforated substance** (20–53).

The **dentate gyrus** (17–53), (8–55), (7–56), also known as the *dentate fascia,* is a narrow, notched band of cortex that lies on the dorsal surface of the hippocampal gyrus and is covered by the **fimbria** (15–53), (7–54), (9–55), (4–56). The anterior portion of the dentate gyrus courses across the medial surface of the uncus and at that point is named the **band of Giacomini** (16–53). Posteriorly the dentate gyrus is continuous with the supracallosal gyrus.

The **hippocampus** (19–53), (8–54), (6–55) is also called *Ammon's horn,* or *cornu ammonis.* It extends along the floor of the inferior horn of the lateral ventricle. The hippocampus is covered by the ependymal lining of the ventricle. A sheet of myelinated fibers known as the **alveus** (6–56) also covers the body of the hippocampus. The fibers of the alveus are continuous with the larger fiber bundle, the **fimbria** (15–53), (7–54).

The area just inferior to the **hippocampal fissure** (8–56) is known as the **prosubiculum** (9–56). The **subiculum** (10–56) lies ventral to it, and the **presubiculum** (11–56) lies along the ventromedial surface of the temporal lobe.

The **fornix** (5–53), (1–54), (4–55) is the principal efferent pathway from the hippocampus. The fibers originate in the hippocampus itself, the hippocampal gyrus, the dentate gyrus, and the prosubiculum. Most of the fibers of the fornix terminate in the homolateral medial nucleus of the **mammillary body** (9–53), (5–54). It does, however, send fibers to the preoptic area, the nucleus of the diagonal band of Broca, the septal area, the haben-

(Continued)

Figure 53

Schematic of the rhinencephalon showing principal structures.

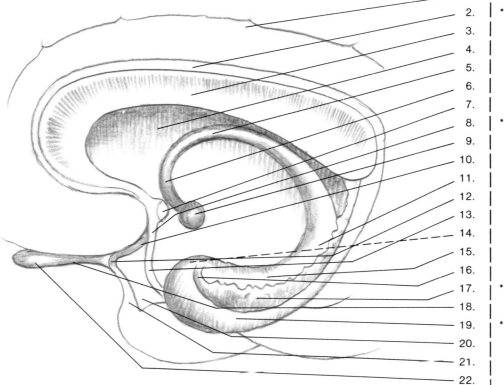

1.	Cingulate gyrus
2.	*Supracallosal gyrus
3.	Corpus callosum
4.	Septum pellucidum
5.	Fornix, body
6.	Fornix, column
7.	Anterior commissure
8.	*Subcallosal gyrus
9.	Mammillary body
10.	Medial olfactory stria
11.	Fornix, crus
12.	Olfactory trigone
13.	Diagonal band of Broca
14.	Uncus
15.	Fimbria
16.	Band of Giacomini
17.	*Dentate gyrus
18.	Olfactory tract
19.	*Hippocampus
20.	Anterior perforated substance
21.	Lateral olfactory stria
22.	Olfactory bulb

ula, and the anterior hypothalamic and subthalamic nuclei. Posteriorly the fornix begins below the splenium of the corpus callosum and arches over the thalamus. The posterior vertical portion of the fornix is the leg or **crus of the fornix** (11–53), (2–54), (10–55). The **body of the fornix** (5–53), (1–54), (4–55) is the portion that lies above the thalamus. It is made up of the joined crura (plural of *crus*). Anteriorly the body divides again and forms the **columns of the fornix** (6–53), (3–54), (3–55), which bend backward behind the anterior commissure and terminate in the mammillary bodies.

The **hippocampal commissure** (4–54), (11–55) is a thin sheet of transverse fibers that runs between the crura of the fornices and establish connections between the hippocampal structures on the two sides of the brain.

Figure 54

Three-dimensional representation of the relationships among the hippocampus, fornix, and mammillary bodies.

1.	Fornix, body
2.	Fornix, crus
3.	Fornix, column
4.	Hippocampal commissure
5.	Mammillary body
6.	Amygdala
7.	Fimbria
8.	*Hippocampus
9.	Uncus

Figure 55

The hippocampus and associated structures in situ.

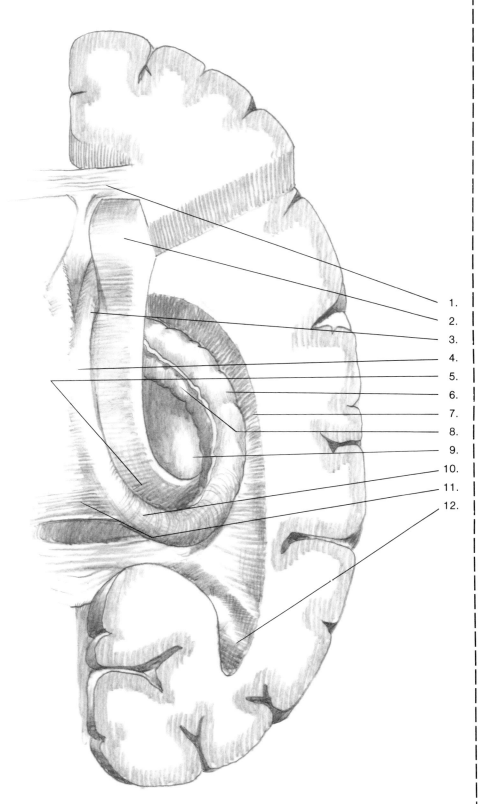

1. Corpus callosum (cut)
2. Caudate nucleus
3. Fornix, column
4. Fornix, body
5. *Thalamus
6. *Hippocampus
7. Lateral ventricle
8. *Dentate gyrus
9. Fimbria
10. Fornix, crus
11. Hippocampal commissure
12. Inferior horn of lateral ventricle

Figure 56

Coronal section through the hippocampus.

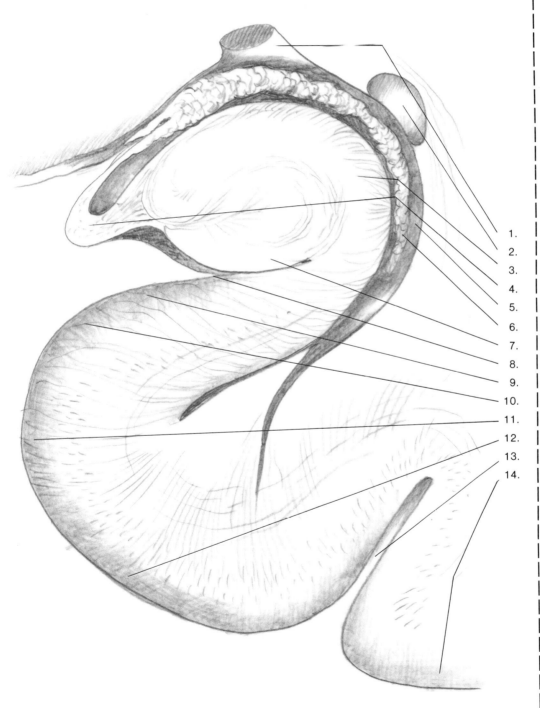

1. Optic tract
2. Caudate nucleus
3. *Hippocampus
4. Fimbria
5. Inferior horn of lateral ventricle
6. Alveus
7. *Dentate gyrus
8. Hippocampal sulcus
9. Prosubiculum
10. Subiculum
11. Presubiculum
12. *Hippocampal gyrus
13. Collateral sulcus
14. *Fusiform gyrus

THE LIMBIC SYSTEM

The limbic system has been proposed as a complex of directly and indirectly connected structures that underlie emotional feelings and thus influence emotional behavior. The *limbic lobe* was originally described by Broca in 1878. It included the cortical areas, which circled the upper brain stem. The relevant structures are the cingulate gyrus, subcallosal gyrus, parahippocampal gyrus, hippocampus, and dentate gyrus. The limbic system as a basis for emotion was initially proposed as a feedback circuit by Papez in 1937. He suggested that central emotive processes were initiated in the hippocampal formation. The impulses were transmitted through the fornix to the mammillary bodies, then to the anterior thalamic nuclei by way of the mammillothalamic tract, and from the anterior thalamic nuclei to the cingulate gyrus.

Since the original proposal by Papez, a number of anatomists have attempted to specify those structures and their connections that are involved in emotional processes. Thus, the limbic system has been variously defined and different anatomists have included different portions of the brain and different pathways. It is generally agreed that this system involves all of the components of the rhinencephalon as described in Section 40, plus a number of other structures and major fiber bundles that connect them.

It is not possible to present all of the structures of the limbic system, the connections between them, and their labels on a single diagram. Figures 57 and 58 present the major components of the limbic system superimposed on the diagram of the rhinencephalon. If Figure 53 has been studied in some detail, the structures will be familiar, and the study of the schematics in Figures 57 and 58 will then provide a good understanding of this complex of interrelated structures.

The limbic system establishes connections among the telencephalon, the diencephalon, and the midbrain tegmentum. The **amygdala** (12–57), which is actually a complex of at least eight separate nuclei, is linked to the **preoptic region** (15–57), (9–58) and the anterior hypothalamic region by the **stria terminalis** (9–57), which also sends fibers to the **septal region** (6–57), (1–58). It should be noted that the principal septal region in humans is located in the subcallosal area and is not the readily recognizable **septum pellucidum** (4–57), which lies between the corpus callosum and the fornix. The septal and the lateral preoptic regions are connected to the **midbrain tegmentum** (8–58) via the **medial forebrain bundle** (11–57), which courses through the lateral hypothalamic area. The tegmental nuclei are also connected to the **mammillary body** (9–53), various nuclei of the hypothalamus, and the septal region by way of the ascending fibers of the **mammillary peduncle** (6–58). The **mammillotegmental tract** (5–58) also connects the midbrain tegmentum and the mammillary bodies.

The **stria medullaris of the thalamus** (2–58), which runs along the border between the dorsal and medial surfaces of the thalamus, establishes connections between the septal area and the **habenula** (10–57), (3–58), where it terminates. From the habenula, connections are made to the **interpeduncular nucleus** (13–57), (7–58) and the tegmental nuclei through the **fasciculus retroflexus** (4–58). The **mammillothalamic tract** (7–57) connects the mammillary body with the anterior nuclear group of the thalamus. The connections between the various thalamic nuclei and the different areas of the cortex have been described in detail earlier (see Sect. 35). Of particular interest here is the projection from the anterior nuclei and the lateral dorsal nucleus of the thalamus to the **cingulate gyrus** (1–57), that is, the **thalamo-cingulate radiation** (5–57). The connection between the dorsomedial nucleus of the thalamus and the **prefrontal cortex** (2–57), the **thalamo-prefrontal radiation** (3–57) is also of interest.

Figure 57

Schematic of some pathways and structures of the limbic system superimposed on the schematic of the rhinencephalon.

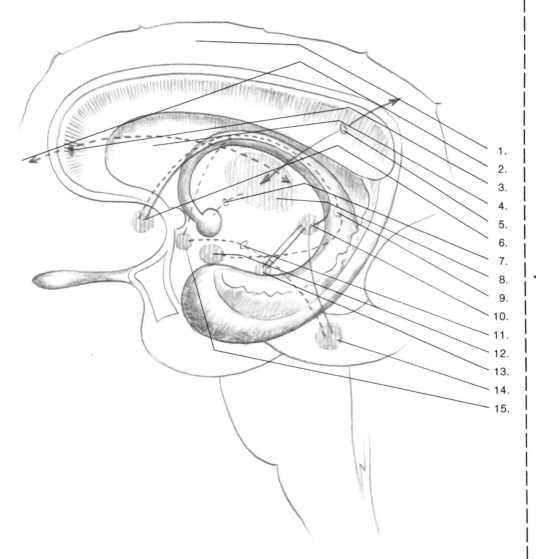

1. Cingulate gyrus
2. Prefrontal cortex
3. Thalamo-prefrontal radiation
4. Septum pellucidum
5. Thalamo-cingulate radiation
6. Septal region
7. Mammillothalamic tract
8. *Thalamus
9. Stria terminalis
10. Habenula
11. Medial forebrain bundle
12. Amygdala
13. Interpeduncular nucleus
14. Tegmental nuclei
15. Preoptic region

Figure 58

Schematic of additional pathways and structures of the limbic system superimposed on the schematic of the rhinencephalon.

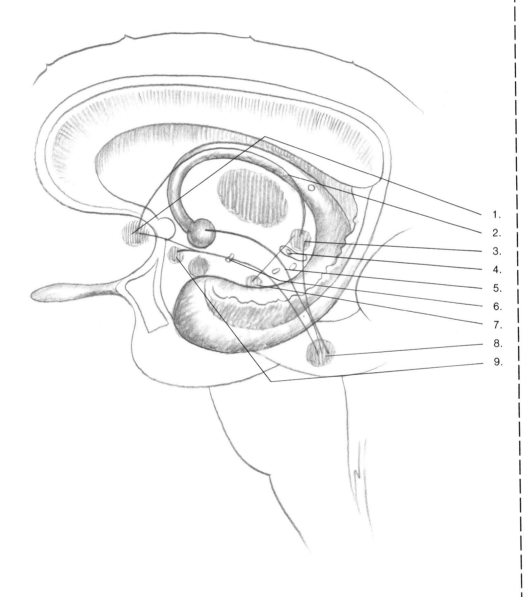

1. Septal region
2. Stria medullaris of thalamus
3. Habenula
4. Fasciculus retroflexus
5. Mammillotegmental tract
6. Mammillary peduncle
7. Interpeduncular nucleus
8. Midbrain tegmentum
9. Preoptic region

THE MENINGES OF THE BRAIN AND SPINAL CORD

The delicate tissue of the brain and the spinal cord is protected by three membranes called **meninges** (singular, **menix**) and the cerebrospinal fluid that circulates among them. Figures 59 and 60 provide a graphic presentation of the meninges and associated structures that will make their relationships to the central nervous system clear. The outermost menix is the **dura mater** (6–59). It is thick, tough, dense, shiny, and inelastic. The dura of the spinal cord is separated from the *periosteum* (the thick fibrous membrane that covers the entire surface of bones) by layers of fat and blood vessels that fill the **epidural space** (12–15). The spinal dura is attached to the second and third cervical vertebrae and around the circumference of the foramen magnum. The spinal dura is a closed sac that runs from the foramen magnum to the second sacral vertebra, where it closely invests the **filum terminale** (4–14) and forms the **coccygeal ligament.** The coccygeal ligament is attached to the dorsal surface of the coccyx.

As the spinal nerves exit from the cord they are invested for a short distance with the dura mater, which is continuous with the connective tissue that surrounds nerve trunks, the epineurium.

Above the foramen magnum the dura mater divides into two layers. The outermost layer is continuous with the periosteum of the vertebral canal and is called the **periosteal dura** (3–59). At certain points the inner layer of the dura, the **meningeal dura mater** (6–59), separates from the periosteal dura and folds on itself projecting into the separations between parts of the brain to form the falx cerebri, the tentorium cerebelli, the falx cerebelli, and the diaphragma sellae.

The **falx cerebri** (7–59), (3–60) is shaped like a sickle and descends vertically into the longitudinal fissure between the cerebral hemispheres. In the front it is quite narrow and is attached to the crista galli of the **ethmoid bone** (9–60). Posteriorly it becomes wider and is continuous with the tentorium cerebelli. The superior border of the falx cerebri forms the floor of the **superior sagittal sinus** (5–59), (2–60). The inferior border arches over the corpus callosum and contains the **inferior sagittal sinus** (4–60).

The **tentorium cerebelli** (13–60) covers the superior surface of the cerebellum and supports the occipital lobes of the brain. Its posterior edge is attached to the inner surface of the occipital bone. The lateral edges are attached to the petrous portion of the temporal bone.

The **falx cerebelli** (6–60) is attached to the tentorium cerebelli and descends into the fissure between the hemispheres of the cerebellum. It is triangular in shape, and its posterior edge is attached to the **occipital bone** (7–60).

The **diaphragma sellae** (10–60) is a circular, horizontal fold of the dura mater that serves to roof the **sella turcica** (11–60), a bony depression into which the pituitary fits. The diaphragma sellae contains a small central hole through which the **infundibulum** (22–29) projects.

The innermost menix is the **pia mater** (12–59), which closely invests the brain and the spinal cord and dips into all of the cerebral sulci and the cerebellar laminae. It is very thin and delicate over the brain but somewhat thicker over the brain stem and spinal cord. The pia mater is invaginated to form the tela choroidea of the third ventricle. It also forms the choroid plexuses of the lateral and third ventricles as well as the tela choroidea and choroid plexuses of the fourth ventricle.

The **arachnoid membrane** (8–59) lies between the dura mater and the pia mater. It is a thin and delicate membrane separated from the pia mater by the **subarachnoid space** (9–59), and from the dura mater by the **subdural space** (11–59), which normally contains only a small amount of moisture. The subarachnoid space contains the cerebrospinal fluid. The arachnoid membrane and the pia mater are connected through the subarachnoid space by delicate fibrous threads called **arachnoid trabeculae** (13–59). These connections give rise to the term *pia-arachnoid,* which is sometimes used to describe them both.

At those points where the contours of the brain do not closely follow the contours of the skull, the subarachnoid space is quite large and contains a significant amount of cerebrospinal fluid. These cavities are called cisterns. The largest is the **cerebellomedullary cistern** (17–62), or *cisterna magna.* Others are the **superior cistern** (12–62), between the cerebellum and the midbrain; the **cisterna pontis** (22–62), just below the pons; the **interpeduncular cistern** (20–62), in the interpeduncular fossa or *posterior perforated substance;* the **cistern of the chiasm** (23–62), which is just anterior to the optic chiasm; and the **cistern of the lateral fossa,** at the lateral cerebral fissure.

The arachnoid membrane projects small, fingerlike projections, called the *arachnoid villi,* into the sagittal sinus. The villi form into complex masses of tissue called **arachnoid granulations** (4–59), or *Pacchionian bodies.*

Figure 59

Coronal section through the superior sagittal sinus showing the meninges.

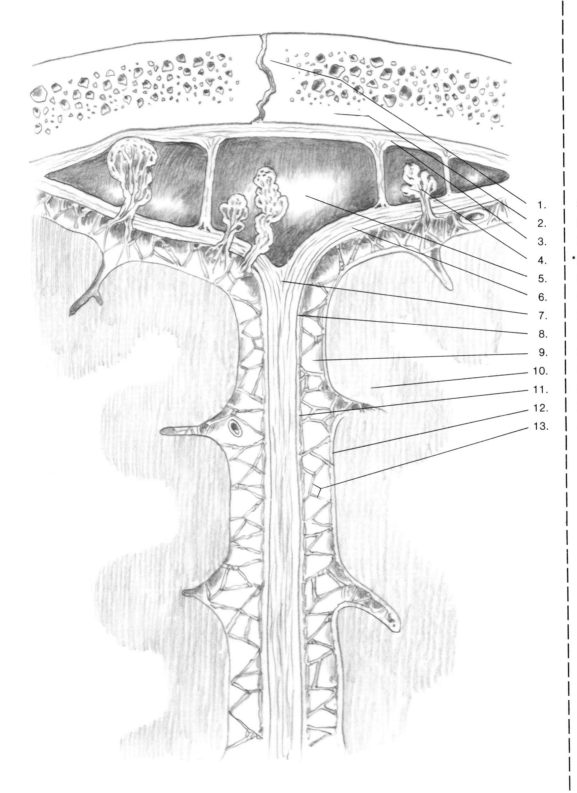

1.   Sagittal suture
2.   Bone
3.   Periosteal dura mater
4.   *Arachnoid granulation
5.   Superior sagittal sinus
6.   Meningeal dura mater
7.   Falx cerebri
8.   Arachnoid
9.   Subarachnoid space
10.   Cerebral cortex
11.   Subdural space
12.   Pia mater
13.   Arachnoid trabeculae

Figure 60

Parasagittal section of the brain showing the cranial dura mater and its major processes.

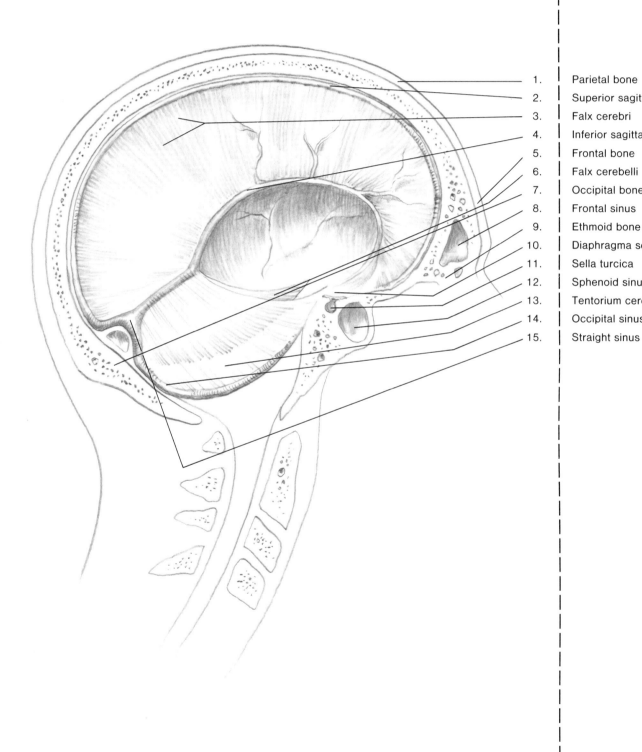

1. Parietal bone
2. Superior sagittal sinus
3. Falx cerebri
4. Inferior sagittal sinus
5. Frontal bone
6. Falx cerebelli
7. Occipital bone
8. Frontal sinus
9. Ethmoid bone
10. Diaphragma sellae
11. Sella turcica
12. Sphenoid sinus
13. Tentorium cerebelli
14. Occipital sinus
15. Straight sinus

THE VENTRICLES OF THE BRAIN

The central nervous system developed from a hollow tube and in maturity retains its original hollow aspect. The spinal canal has been discussed earlier. It runs the entire length of the spinal cord and extends into the medulla oblongata through the closed portion of the medulla. It then expands (see Fig. 61) into the **fourth ventricle** (13–61), the floor of which is the **rhomboid fossa** (33–8) and the roof, the cerebellum. Anteriorly it is continuous with the **cerebral aqueduct** (11–61) of the midbrain. The lateral portion of the fourth ventricle extends into the **lateral recess** which has an opening at its end. This opening is called the **lateral aperture** (16–61), or *foramen of Luschka.* Near the caudal extremity of the roof of the fourth ventricle is another opening, the **medial aperture** (12–61), or *foramen of Magendie.* The **choroid plexus of the fourth ventricle** (16–62) consists of a mass of small blood vessels primarily of capillary size. They are derived from the pia mater and are covered with the epithelial ependymal lining of the ventricle. The choroid plexus secretes cerebrospinal fluid into the ventricle. The plexus is located in the roof of the ventricle and extends out through the lateral recess, and a small amount protrudes through the lateral apertures. The three apertures of the fourth ventricle provide channels that connect the ventricle with the **subarachnoid space** (9–59).

The **cerebral aqueduct** (11–61), or *aqueduct of Sylvius,* opens into the **third ventricle** (5–61), which is surrounded by the **diencephalon** (3–2). The third ventricle is a narrow cleft, the sides of which are formed by the two thalami. Its anterior boundary is the **anterior commissure** (17–10) and the *lamina terminalis,* which is a thin plate running between the optic chiasm and the anterior commissure joining the two hemispheres. In the rostral boundary of the third ventricle is a protrusion known as the **optic recess** (17–61). The **infundibular recess** (14–61) lies inferior to the optic recess. The posterior aspect of the ventricle also contains two recesses, the **suprapineal recess** (6–61) and the **pineal recess** (10–61). The **intermediate mass** (7–61), which connects the two thalami, penetrates the medial portion of the third ventricle. The roof of the third ventricle also contains a choroid plexus that secretes cerebrospinal fluid. The third ventricle is connected to the lateral ventricles by the **interventricular foramen** (4–61), also called the *foramen of Monro.*

The two lateral ventricles each have a central part or body, an anterior horn, a posterior horn, and an inferior horn. The **anterior horn** (3–61) lies rostral to the interventricular foramen. A coronal section of this horn reveals that it has a triangular shape. The **septum pellucidum** (3–10), which is a relatively thin tissue, serves as the medial aspect of the anterior horn and separates the two lateral ventricles. The **corpus callosum** (1–10) serves as the roof and the anterior boundary of the anterior horn. The floor and the lateral wall are formed by the **head of the caudate nucleus** (2–40).

The corpus callosum also forms the roof of the **body of the lateral ventricle** (1,2–61), which extends back to the splenium of the corpus callosum, at which point there is a junction between the **posterior horn** (8–61) and the **inferior horn** (15–61). The enlargement produced by this junction is referred to as the **collateral trigone** (9–61). The floor of the body of the lateral ventricle is composed of a number of different structures, including the choroid plexus of the lateral ventricle, parts of the thalamus and the caudate nucleus, as well as the **fornix** (1–34a), and **stria terminalis** (2–34c).

The **posterior horn** (8–61) projects into the occipital lobe and tapers to a point. The roof and lateral walls are formed by the tapetum of the corpus callosum.

The inferior horn curves ventrally and then in a rostral direction into the temporal lobe almost to the temporal pole. It is the largest of the three horns of the lateral ventricles. The roof of the inferior horn is formed by the inferior surface of the tapetum of the corpus callosum. The floor of the inferior horn includes the **hippocampus** (19–53) and the **fimbria** (15–53).

The choroid plexus of the lateral ventricle lies on the floor of the ventricle and continues through the interventricular foramen to become continuous with the choroid plexus of the third ventricle. It also extends into the inferior horn almost to its rostral end.

Figure 61

The ventricles of the brain.

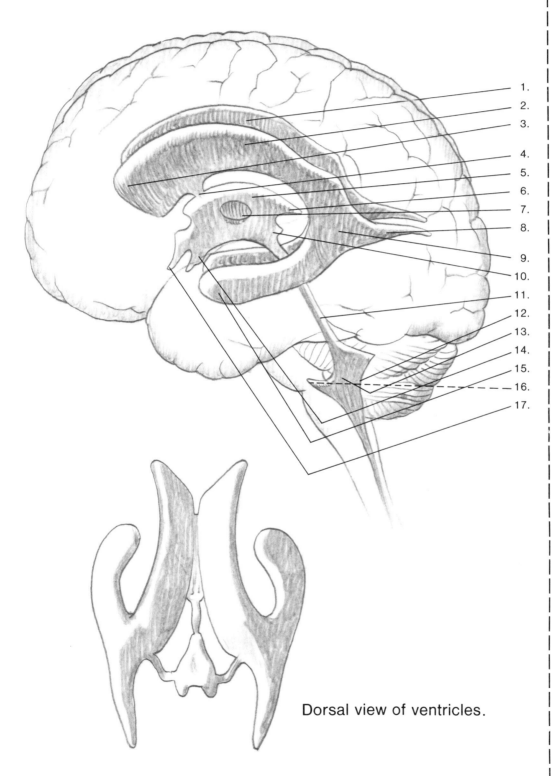

1. Lateral ventricle, right body
2. Lateral ventricle, left body
3. Anterior horn of lateral ventricle
4. *Interventricular foramen
5. Third ventricle
6. Suprapineal recess
7. *Intermediate mass
8. Posterior horn of lateral ventricle
9. Collateral trigone
10. Pineal recess
11. *Cerebral aqueduct
12. *Medial aperture
13. Fourth ventricle
14. Infundibular recess
15. Inferior horn of lateral ventricle
16. *Lateral aperture
17. Optic recess

Dorsal view of ventricles.

THE CIRCULATION OF THE CEREBROSPINAL FLUID

The cerebrospinal fluid is a clear, colorless, watery material similar in composition to that produced by the lymph glands. It is contained in the brain ventricles, the subarachnoid spaces, and in the central canal of the spinal cord. The system contains between 125 and 150 milliliters. It functions as an hydraulic buffer to protect the delicate tissue of the central nervous system from trauma. It also provides a fluid vehicle for the transmission of chemical substances to the intercellular spaces of the brain (Fig. 62).

The cerebrospinal fluid is formed primarily in the **choroid plexuses** (9,10,16–62) of the ventricles. There is evidence, however, that it is also formed in the central canal. The fluid circulates slowly throughout the system. There is no connection between the **lateral ventricles** (8–62) and the subarachnoid space, so the fluid produced in the lateral ventricles leaves through the **interventricular foramen** (19–62), or *foramen of Monro,* which communicates with the **third ventricle** (11–62). From there it passes through the **cerebral aqueduct** (14–62), or *aqueduct of Sylvius,* into the **fourth ventricle** (15–62). From the fourth ventricle it enters the subarachnoid space through the three apertures of that ventricle. The two **lateral apertures,** or *foramen of Luschka,* empty into the **cisterna pontis** (22–62), while the **medial aperture** (18–62), or *foramen of Magendie,* opens into the **cerebellomedullary cistern** (17–62), or *cisterna magna.* From the cerebellomedullary cistern, the fluid circulates rostrally around the cerebellum to the **superior cistern** (12–62) and into the subarachnoid space around the hemispheres. It also circulates caudally into the subarachnoid space around the cord. From the cisterna pontis the flow is toward the **interpeduncular cistern** (20–62) and the **cistern of the chiasm** (23–62). From here it proceeds rostrally in the subarachnoid space around the cerebral hemispheres and into the longitudinal fissure and the lateral fissures. Ultimately it is absorbed by the arachnoid villi in the **arachnoid granulations** (2–62) and empties into the great venous sinuses of the dura, especially the **superior sagittal sinus** (1–62).

Figure 62

The circulation of the cerebrospinal fluid.

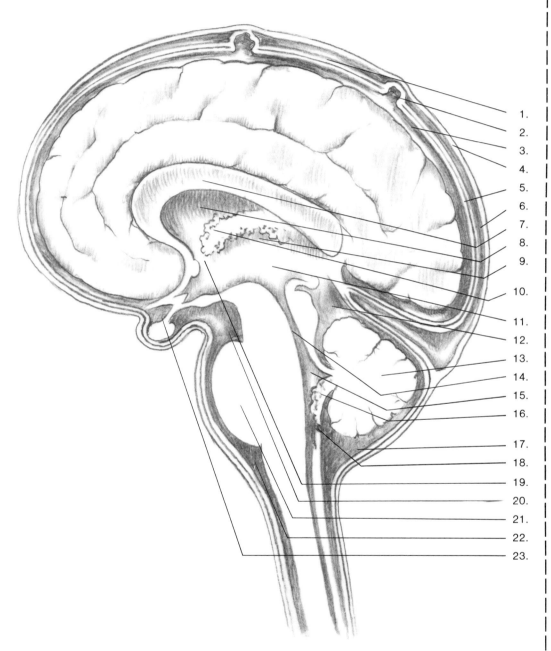

1. Superior sagittal sinus
2. *Arachnoid granulation
3. Subarachnoid space
4. Periosteal dura mater
5. Arachnoid
6. Meningeal dura mater
7. Corpus callosum
8. Lateral ventricle
9. Choroid plexus of third ventricle
10. Choroid plexus of lateral ventricle
11. Third ventricle
12. Superior cistern
13. Cerebellum
14. *Cerebral aqueduct
15. Fourth ventricle
16. Choroid plexus of fourth ventricle
17. *Cerebellomedullary cistern
18. *Medial aperture
19. *Interventricular foramen
20. Interpeduncular cistern
21. Pons
22. Cisterna pontis
23. Cistern of chiasm

THE CRANIAL NERVES AND THEIR NUCLEI

In order to facilitate study of the cranial nerves four figures and two tables are presented. Two sagittal (Figs. 63 and 64) and one dorsal (Fig. 65) view of the nuclei of the cranial nerves are given, and a ventral view of the brain stem (Fig. 66) is shown in which most of the detail has been eliminated except for the exit points of the cranial nerves. Familiarity with the cranial nerves is essential. The tables have been prepared so that the fundamental information can be more readily learned. In Table 2, the first cranial nerve table, the number of the nerve is given in the left column. The name of the nerve is in the next column, and the central connection is given in the third. In Table 3, the second cranial nerve table, the number is given in the left column with the function and the peripheral distribution in the second and third. After adequate study, self-testing can be accomplished by covering columns two and three with a blank sheet of paper and writing in the appropriate facts just as the structures are identified on the columns of labels on the diagrams.

There are twelve pairs of cranial nerves, and each has been assigned a Roman numeral. Since the cranial nerves are frequently referred to by a number rather than by name, it is imperative that you associate the names and the numbers. Generations of students in the biological sciences have been using mnemonic devices to remember the cranial nerves in order. One of the easiest devices for remembering them is as follows:

On	old	Olympus'	towering	top	a
I	II	III	IV	V	VI

fat	-	assed	German	vaults	and	hops.
VII		VIII	IX	X	XI	XII

The first letter of each word in the sentence is the same as the first letter in the indicated cranial nerve.

Most of the cranial nerves, like the spinal nerves, contain both sensory and motor components and are *mixed nerves*. A few, however, have only a sensory or a motor component and may be considered *pure nerves*. The sensory, or afferent, components of the cranial nerves have their cell bodies outside the brain, while the cell bodies of the motor components are located in the cranial nerve nuclei within the brain.

TABLE 2

THE CENTRAL CONNECTION OF THE CRANIAL NERVES		
NERVE NUMBER	**NAME**	**CENTRAL CONNECTION**
I	olfactory	olfactory bulb and tract
II	optic	optic nerve and tract
III	oculomotor	nucleus of the oculomotor nerve Edinger-Westphal nucleus
IV	trochlear	nucleus of the trochlear nerve
V	trigeminal	sensory nucleus of the trigeminal nerve motor nucleus of the trigeminal nerve
VI	abducens	nucleus of the abducens nerve
VII	facial	nucleus of the facial nerve superior salivatory nucleus nucleus of the tractus solitarius
VIII	acoustic	cochlear nuclei vestibular nuclei
IX	glossopharyngeal	nucleus ambiguus inferior salivatory nucleus nucleus of the tractus solitarius
X	vagus	nucleus of the tractus solitarius nucleus ambiguus dorsal motor nucleus of the vagus nerve
XI	accessory	nucleus of the accessory nerve nucleus ambiguus dorsal motor nucleus of the vagus nerve
XII	hypoglossal	hypoglossal nucleus

(Continued)

Figure 63

Sagittal view of the nuclei and motor components of the cranial nerves.

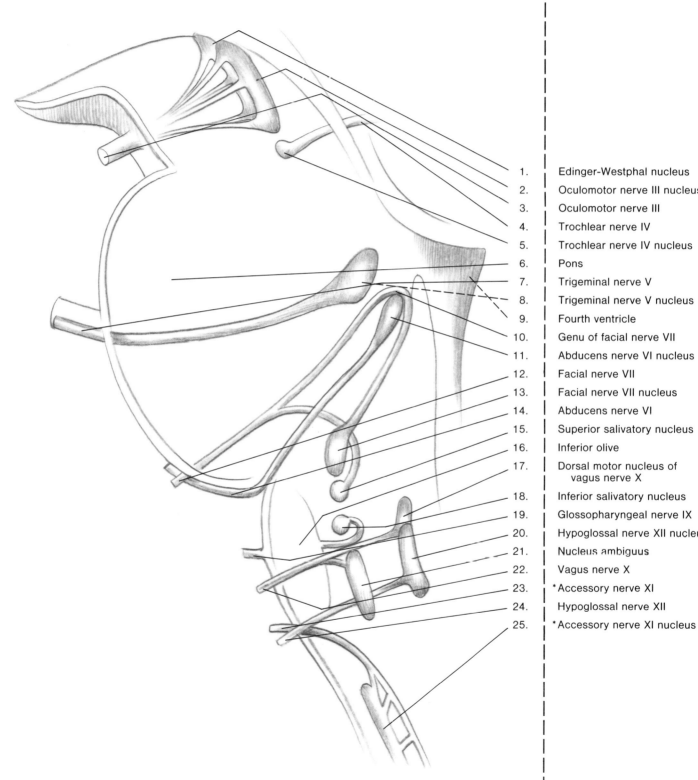

1. Edinger-Westphal nucleus
2. Oculomotor nerve III nucleus
3. Oculomotor nerve III
4. Trochlear nerve IV
5. Trochlear nerve IV nucleus
6. Pons
7. Trigeminal nerve V
8. Trigeminal nerve V nucleus
9. Fourth ventricle
10. Genu of facial nerve VII
11. Abducens nerve VI nucleus
12. Facial nerve VII
13. Facial nerve VII nucleus
14. Abducens nerve VI
15. Superior salivatory nucleus
16. Inferior olive
17. Dorsal motor nucleus of vagus nerve X
18. Inferior salivatory nucleus
19. Glossopharyngeal nerve IX
20. Hypoglossal nerve XII nucleus
21. Nucleus ambiguus
22. Vagus nerve X
23. *Accessory nerve XI
24. Hypoglossal nerve XII
25. *Accessory nerve XI nucleus

The **olfactory nerve [I]** (14–77) is a purely sensory nerve and serves the sense of smell. It originates in the olfactory cells of the nasal mucosa. Bundles of fibers collect and pass through the cribriform plate of the ethmoid bone as the olfactory nerve [I] and end in the **olfactory bulb** (1–66), which continues in a caudal direction as the **olfactory tract** (2–66). The bulb and tract appear to be the cranial nerve but are actually a part of the rhinencephalon, as discussed in Section 40. The olfactory tract divides into the **lateral and medial stria** (10–53), (21–53). The medial branch ends in the subcallosal gyrus, while the lateral branch ends in the uncus and hippocampal gyrus.

The **optic nerve [II]** (3–66) is also a pure sensory nerve serving the sense of vision. It originates in the ganglion cells of the retina. The axons of these cells pass through the optic foramen and proceed to the **optic chiasm** (5–9), where the fibers from the medial half of each retina cross to the opposite side while those from the lateral half of the retina remain on the same side. This mixture of fibers from both eyes passes caudalward in the optic tract to the lateral geniculate body. In the geniculate body the fibers of the optic tract synapse with the secondary neurons of the geniculocalcarine tract, which continues to the calcarine cortex of the occipital lobe.

The **oculomotor nerve [III]** (3–63), (1–65), (4–66) is a mixed nerve that serves to control all of the eye muscles except the lateral rectus and the superior oblique. Its afferent component provides a muscle sense for the muscles it controls. The oculomotor nerve [III] also contains parasympathetic fibers that control the ciliary muscle and the sphincter of the pupil. It exits from the inferior

TABLE 3

NERVE NUMBER	FUNCTION	PERIPHERAL DISTRIBUTION
	THE FUNCTION AND PERIPHERAL DISTRIBUTION OF THE CRANIAL NERVES	
I	smell	olfactory nerves
II	vision	retina, rods and cones
III	eye movement	all eye muscles except lateral rectus and superior oblique
	eye muscle sensation	sensory endings in eye muscles
	pupil contraction	ciliary muscle and sphincter of the pupil
IV	eye movement	superior oblique muscles of eye
	eye muscle sensation	sensory endings of superior oblique
V	sensation from skin and mucous membranes of head and from mastication muscles	sensory branches of opthalmic, maxillary, and mandibular nerves
	chewing movements	muscles of mastication
VI	eye movement	lateral rectus muscle
	eye muscle sensation	sensory endings in lateral rectus
VII	facial expression	voluntary facial muscles
	glandular secretion	glands of nasal mucosa, sublingual and submaxillary glands, lacrimal glands
	taste	taste buds, anterior ⅔ of tongue
VIII	hearing	organ of Corti in cochlea
	sense of equilibrium	semicircular canals, saccule, and utricle
IX	taste	taste buds of posterior ⅔ of tongue
	sensation from tongue and throat	branches of tympanic nerve to tongue, pharynx, and larynx
	swallowing	throat muscles
	glandular secretion	parotid gland
X	sensation from throat, viscera	larynx, pharynx, esophagus, heart, respiratory tract, most of digestive tract
	parasympathetic control of viscera	heart, lungs, digestive tract, and most of the viscera
XI	swallowing and phonation	muscles of pharynx, larynx, and soft palate
	head and shoulder movement	sternocleidomastoid and trapezius muscles
XII	tongue movements	extrinsic and intrinsic tongue muscles

(Continued)

Figure 64

Sagittal view of the nuclei and sensory components of the cranial nerves.

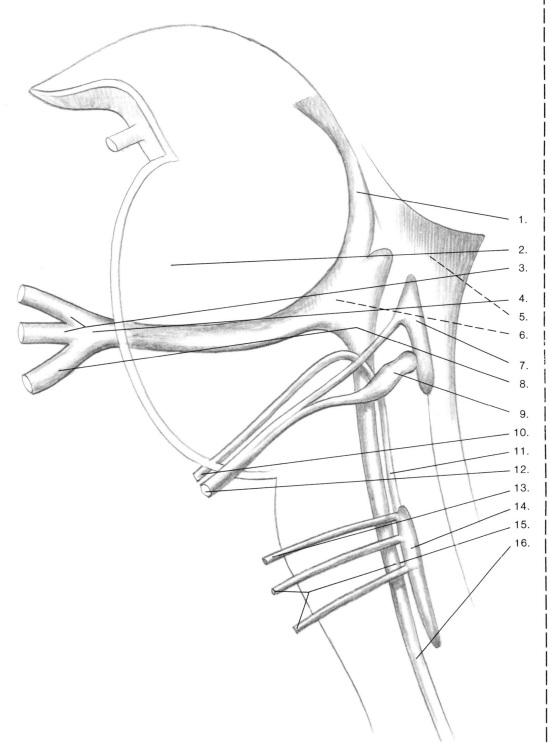

1.   Mesencephalic tract of trigeminal nerve V
2.   Pons
3.   Trigeminal nerve V, sensory branches
4.   *Semilunar ganglion
5.   Fourth ventricle
6.   Sensory nucleus of the trigeminal nerve V
7.   Vestibular nucleus
8.   Trigeminal nerve V, motor branch
9.   Cochlear nucleus
10. *Intermediate nerve VII
11. Tractus solitarius
12. *Acoustic nerve VIII
13. Glossopharyngeal nerve IX
14. Nucleus of tractus solitarius
15. Vagus nerve X
16. Spinal tract of trigeminal nerve V

surface of the midbrain just anterior to the pons, travels in a rostral direction, and enters the orbit through the superior orbital fissure. The oculomotor nerve [III] has a superior and an inferior branch. The superior branch supplies the superior rectus and levator palpebrae muscles, while the inferior branch goes to the medial and inferior recti and the inferior oblique muscles. The inferior branch also sends fibers to the ciliary ganglion. The **nucleus of the oculomotor nerve [III]** (2–63), (3–65) is located in the tegmentum of the mesencephalon below the colliculi. The oculomotor nerve [III] also receives fibers from the **Edinger–Westphal nucleus** (1–63), (2–65) located just superior to the oculomotor nucleus.

The **trochlear nerve [IV]** (4–63), (4–65), (9–66) is a small mixed nerve that sends motor fibers to the superior oblique muscles of the eye. It also includes afferent fibers from the same muscles. It exits from the brain stem just behind the inferior colliculus of the mesencephalon and then curves around the lateral surface of the cerebral peduncle at a point just above the pons. The trochlear nerve [IV] crosses over the oculomotor nerve [III] before entering the superior orbital fissure. The **nucleus of the trochlear nerve [IV]** (5–63), (5–65) lies just caudal to the nucleus of the oculomotor nerve [III].

The **trigeminal nerve [V]** (7–63) is a large mixed nerve. The sensory division carries impulses from touch, pain, heat, and cold receptors of the facial area, the scalp, and the mucous membranes. It also contains some proprioceptive fibers and some visceral efferents of the head region.

The sensory division is divided into three branches. The *opthalamic branch* receives fibers from the eyeball, the conjunctiva, the cornea, and the lacrimal gland. It also receives sensory fibers from the skin of the anterior half of the scalp, including the forehead as well as the side of the nose, the eyelid, and the eyebrow. The *maxillary branch* receives sensory fibers from the skin of the cheek and from the mucous membrane of the nose, the palate, and adjacent parts of the pharynx. It also has branches from the upper teeth. The *mandibular branch* includes sensory fibers from the lower teeth, the skin of the face over the mandible, and the side of the head in front of the ear. It also receives fibers from the front two-thirds of the tongue and the floor of the mouth.

The **motor branch of the trigeminal nerve [V]** (8–64), (6–65) is relatively small. It runs along with the mandibular branch of the sensory division and innervates the muscles of mastication. The **motor nucleus of the trigeminal nerve [V]** (8–65) is small and is in the tegmentum of the pons. The **sensory nucleus of the trigeminal nerve [V]** (6–64) is located in the mesencephalon and in the pons and extends as far down as the cervical section of the spinal cord. The nerve emerges from the lateral portion of the pons in two roots. The large sensory root has a significant ganglion called the **semilunar ganglion** (4–64), (8–66), or *Gasserian ganglion* from which the three sensory branches of the trigeminal nerve [V] emerge. The opthalamic branch passes through the superior orbital fissure, the maxillary branch goes through the foramen rotundum, and the manibular branch enters the foramen ovale.

The **abducens nerve [VI]** (14–63), (10–65), (10–66) is a mixed nerve that contributes to the control of eye movement and mediates muscle sense. It supplies both afferent and efferent fibers to the lateral rectus muscles of the eyes. It emerges from the brain stem at the inferior border of the pons, passes along the cavernous sinus, and enters the orbit by way of the superior orbital fissure. The **nucleus of the abducens nerve [VI]** (12–65) is located in the posterior portion of the tegmentum of the pons.

The **facial nerve [VII]** (12–63), (9–65), (11–66) is a mixed nerve that innervates the muscles of the scalp and face. It also contains some visceral efferent fibers that supply the lacrimal glands and the sublingual and submaxillary glands. The facial nerve [VII] also supplies glands of the mouth, nose, pharynx, and palate. It reaches the facial muscles after passing through the parotid gland.

The sensory part of the facial nerve [VII] is distinguishable on the ventral aspect of the brain stem and is given the name **intermediate nerve [VII]** (10–64), (16–65), (12–66). It receives taste fibers from the anterior two-thirds of the tongue and parts of the palate. The nerve from the tongue is the chorda tympani. It runs part way with the lingual nerve from the trigeminal and joins the facial nerve [VII] later.

The facial nerve [VII] originates in the **nucleus of the facial nerve [VII]** (13–63), (14–65) located in the tegmentum of the inferior portion of the pons ventral and lateral to the nucleus of the abducens. The fibers course in a dorsomedial direction, go around the nucleus of the abducens nerve [VI], and emerge from the brain at the caudal border of the pons. It travels through the internal auditory meatus, goes through the facial canal of the temporal bone, and exits through the stylomastoid fora-

(Continued)

Figure 65

Dorsal view of the cranial nerves, with sensory components on the left and motor components on the right.

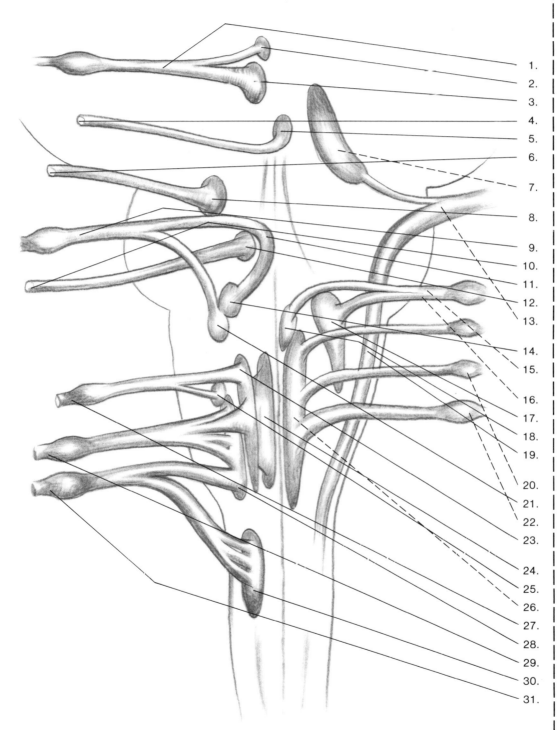

1. Oculomotor nerve III
2. Edinger-Westphal nucleus
3. Oculomotor nerve III nucleus
4. Trochlear nerve IV
5. Trochlear nerve IV nucleus
6. Trigeminal nerve V, motor component
7. Mesencephalic nucleus of trigeminal nerve V
8. Motor nucleus of trigeminal nerve V
9. Facial nerve VII
10. Abducens nerve VI
11. Genu of facial nerve VII
12. Abducens nerve VI nucleus
13. Sensory components of trigeminal nerve V
14. Facial nerve VII nucleus
15. *Acoustic nerve VIII, auditory branch
16. *Intermediate nerve VII
17. Vestibular nucleus
18. Cochlear nucleus
19. Spinal tract of trigeminal nerve V
20. Glossopharyngeal nerve IX
21. Superior salivatory nucleus
22. Vagus nerve X
23. Dorsal motor nucleus of vagus nerve X
24. Hypoglossal nerve XII nucleus
25. Inferior salivatory nucleus
26. Tractus solitarius
27. Glossopharyngeal nerve IX
28. Nucleus ambiguus
29. Vagus nerve X
30. *Accessory nerve XI nucleus
31. *Accessory nerve XI

men. It reaches the facial muscles after passing through the parotid gland. Some of the fibers of the facial nerve [VII] originate in the **superior salivatory nucleus** (15–63), (21–65), and others originate in the **nucleus of the tractus solitarius** (14–64).

The **acoustic nerve [VIII]** (12–64), (15–65), (13–66) is also called the *statoacoustic nerve* or *vestibulocochlear nerve*. It consists of two purely sensory sections, one for audition and one for the vestibular sense. The *acoustic divisino* of the eighth cranial nerve is related to hearing. It arises in the inner ear from the *spiral organ of Corti* with its cell bodies located in the spiral ganglion. The fibers pass through the internal acoustic meatus and terminate in the *cochlear nucleus* (9–64), (18–65). Fibers cross in the lateral lemniscus from the ventral portion of the cochlear nucleus to the inferior colliculus. Fibers from the dorsal part of the nucleus go through the medial geniculate body to the inferior colliculus and then to the medial portion of the temporal lobe of the cortex.

The *vestibular division* of the eighth nerve is concerned with equilibrium and originates in the cells of the utricle and saccule and the semicircular canals. It proceeds to the vestibular ganglion in the temporal bone and goes from there to the **vestibular nucleus** (7–64), (17–65) in the medulla.

The **glossopharyngeal nerve [IX]** (13–64), (27–65), (14–66) is a mixed nerve. The sensory component involves taste, and the motor portion provides for reflex control of blood pressure and swallowing. It emerges from the rostral portion of the medulla and exits from the skull through the jugular foramen with the vagus [X] and accessory [XI] nerves. It supplies sensation to the posterior third of the tongue and to the mucous membrane of the pharynx. These sensory fibers terminate in the rostral part of the **nucleus of the tractus solitarius** (14–64). Some efferent fibers originate in the anterior portion of the **nucleus ambiguus** (21–63), (28–65) and go to the stylopharyngeus muscle and contribute to the control of swallowing. This nerve also contains preganglionic parasympathetic fibers that go to the otic ganglion. The postganglionic fibers proceed from the otic ganglion to the parotid gland. These fibers arise in the **inferior salivatory nucleus** (18–63), (25–65).

The **vagus nerve [X]** (22–63), (15–64), (29–65), (15–66) is a mixed nerve that carries visceral afferent fibers from the heart, digestive tract, and most of the rest of the viscera. It also includes sensory fibers from the larynx, pharynx, and the esophagus. The motor component includes parasympathetic fibers to all of these structures. It originates in a series of rootlets emanating from the lateral portion of the medulla with the glossopharyngeal nerve [IX] above it and the accessory nerve [XI] below. It exits from the cranial cavity through the jugular foramen. Fibers of the vagus nerve [X] terminate in the nucleus of the tractus solitarius. Motor fibers from the vagus nerve [X] originate in the nucleus ambiguous and the **dorsal motor nucleus of the vagus nerve [X]** (17–63), (23–65).

The **accessory nerve [XI]** (23–63), (31–65), (17–66), also called the *spinal accessory nerve,* is a motor nerve. It is divided into a spinal and a cranial section. The spinal section is derived from the rostral five segments of the spinal cord. They originate in the **nucleus of the accessory nerve [XI]** (25–63), (30–65), and form a nerve trunk that enters the cranial cavity through the foramen magnum to join the cranial section of the accessory nerve [XI]. They both exit from the jugular foramen. The spinal fibers then proceed downward to supply the sternocleidomastoid and the trapezius muscles. The cranial fibers of the accessory nerve [XI] originate in the **nucleus ambiguus** and the **dorsal motor nucleus of the vagus nerve [X].**

Superficially the cranial section of the accessory nerve [XI] originates from a position on the lateral medulla just below the vagus nerve [X]. It supplies the voluntary muscles of the pharynx, larynx, and soft palate.

The **hypoglossal nerve [XII]** (24–63), (16–66) is a motor nerve that supplies the extrinsic and intrinsic muscles of the tongue. It originates in the **hypoglossal nucleus** (20–63), (24–65), leaves the brain stem as a series of rootlets on the ventral aspect, and passes through the hypoglossal canal.

Figure 66

Ventral view of the exits of the cranial nerves.

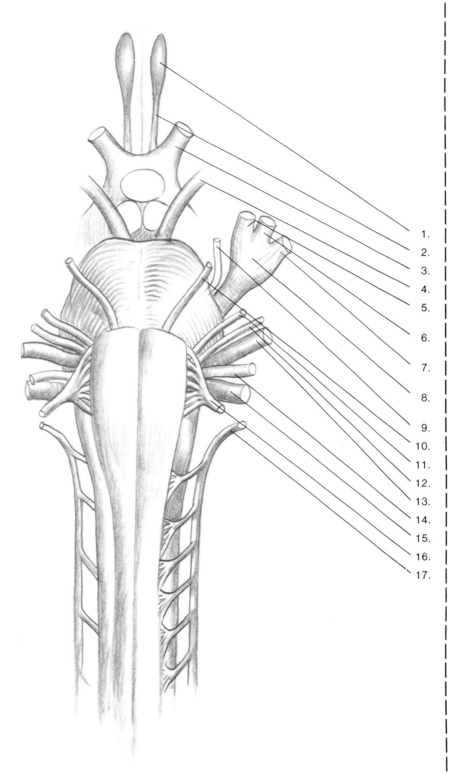

1. Olfactory bulb
2. Olfactory tract
3. Optic nerve II
4. Oculomotor nerve III
5. Opthalamic branch of trigeminal nerve V
6. Maxillary branch of trigeminal nerve V
7. Mandibular branch of trigeminal nerve V
8. *Semilunar ganglion of trigeminal nerve V
9. Trochlear nerve IV
10. Abducens nerve VI
11. Facial nerve VII
12. *Intermediate nerve VII
13. *Acoustic nerve VIII
14. Glossopharyngeal nerve IX
15. Vagus nerve X
16. Hypoglossal nerve XII
17. *Accessory nerve XI

THE AUTONOMIC NERVOUS SYSTEM

The autonomic nervous system includes those nerve fibers that supply motor control of the glands and of the smooth and cardiac muscles of the body. It is a complex system involving all parts of the body. As a result, diagrams of the autonomic system become so complex that they are extremely difficult to follow. On the other hand, all of the parts of the autonomic nervous system are interrelated, so it is difficult to diagram one part without showing its relationship to another part. Figure 67 shows the autonomic reflexes and the relationship between the sympathetic and the somatic nervous systems. Figures 68 and 69 show the sympathetic and the parasympathetic nervous systems and their relationship to each other. The diagram of the sympathetic system has the parasympathetic drawn in lightly, and the diagram of the parasympathetic system also has the sympathetic indicated.

Most anatomists agree that only motor nerves should be included under the autonomic classification. This, of course, is not to imply that the viscera and other structures supplied by the autonomic system do not have sensory input to the central nervous system. In fact, they do. Further, the visceral afferents run along in the same nerve bundles. However, one characteristic of the autonomic nervous system is not shared by the visceral afferents. All of the nerves of the autonomic system consist of two sets of fibers that are connected at a synapse outside the central nervous system. This is not true of the visceral afferents. The cell bodies of those fibers are located in the spinal ganglia of the posterior nerve root, as are the cell bodies of all of the other afferent or sensory fibers of the nervous system. Figure 67 makes that clear. Although the **visceral afferent neurons** (10–67) shown in the diagram do pass through the sympathetic ganglion, they do not synapse there.

The relationship between the sympathetic and the somatic nervous systems can best be seen in Figure 67. The cell bodies for the sympathetic system lie in the lateral horn of the spinal-cord gray matter. They are distributed from the first thoracic to the third lumbar segments, which is why the sympathetic is also referred to as the *thoracolumbar* system. The sympathetic neuron leaves the cord through the **ventral root** (7–67) and travels out to a point just beyond the chains of **sympathetic ganglia** (15–67). At this point it curves medially and enters the sympathetic ganglion through the **white ramus communicans** (8–67). The nerve that originates in the lateral horn of the cord is a **preganglionic neuron**

(12–67), (1–69) because it has not yet made a synapse. Preganglionic neurons are myelinated, with the result that the ramus communicans to the ganglion is white in appearance.

After the preganglionic neuron enters the sympathetic trunk, it may take several possible courses. First, it may synapse in the ganglion it enters with the unmyelinated **postganglionic neuron** (9–67), (2–69). Secondly, it may proceed either up or down the sympathetic chain and synapse with a postganglionic neuron in a ganglion above or below its point of entry. Third, it may proceed through the sympathetic trunk without synapsing and enter one of the collateral ganglia, such as the **celiac ganglion** (13–67), (15–68), where it makes connections with the secondary neuron. Finally, it may follow any combination of these courses.

The sympathetic neurons that serve the blood vessels and glands of the skin are postganglionic and leave the sympathetic trunk and rejoin the peripheral nerve. The nerve bundle between the ganglion and the peripheral nerve is the **gray ramus communicans** (6–67). It is gray because the postganglionic neurons are unmyelinated.

As Figure 67 shows, the visceral afferent neurons do not synapse outside the central nervous system. They run from the organ of innervation directly to the gray matter of the spinal cord, where they synapse. Their cell bodies lie in the **spinal ganglia** (2–67). The visceral afferents synapse with the preganglionic cells of the sympathetic system, thus providing for sympathetic reflexes.

Figure 67

Autonomic reflexes and the relationship between the sympathetic and the somatic nervous systems.

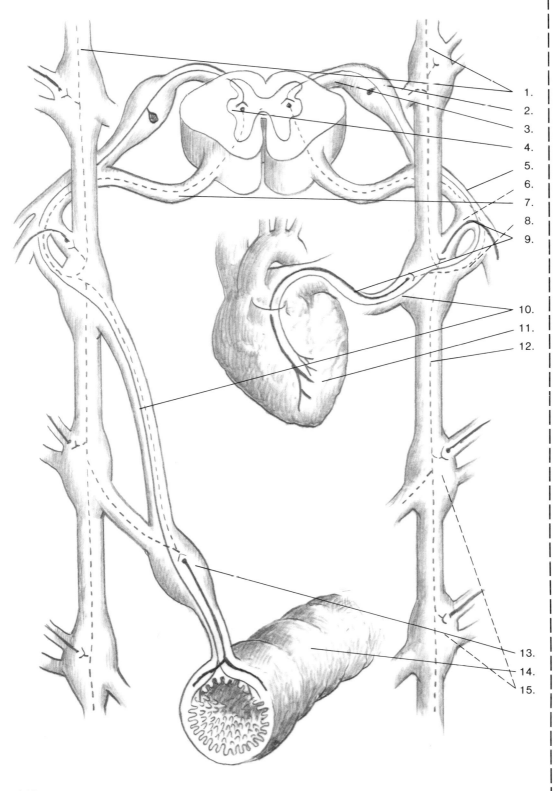

1. Sympathetic chain
2. Spinal ganglion
3. Dorsal root
4. *Lateral horn
5. Spinal nerve
6. Gray ramus communicans
7. Ventral root
8. White ramus communicans
9. Postganglionic neurons

10. Visceral afferent neurons
11. Heart
12. Preganglionic neurons

13. Celiac ganglion
14. Intestine
15. Sympathetic ganglia

THE SYMPATHETIC DIVISION OF THE AUTONOMIC NERVOUS SYSTEM

The sympathetic division of the autonomic nervous system sends neurons to smooth muscles and glands throughout the body (see Fig. 68). In general it functions to place the individual in a state that constitutes preparation for an emergency situation. For example, activation of the sympathetic system results in an increase in heart rate, a dilation of the pupils, a constriction of the blood vessels of the digestive tract, an increase in blood pressure, the secretion of adrenalin from the adrenal medulla, and the breakdown of glycogen in the liver.

Most of the same muscles and glands receive fibers from the parasympathetic nervous system. In general, although not in every case, the sympathetic and parasympathetic divisions act in opposition, with the result that the body is in a state of balance with neither system dominating.

Action of the parasympathetic system results in bodily changes more characteristic of calm and nonstressful states. The pupils of the eyes constrict, the heart rate slows, the blood pressure drops, and the digestive system becomes active and functioning.

Although external stress generally produces an activation of the sympathetic system and an inhibition of the parasympathetic, if the onset of the sympathetic activity is sudden and intense there will be an intensive parasympathetic compensatory action. For example, sympathetic action produces a contraction of the anal-sphincter muscles, and parasympathetic action produces an inhibition to the contraction of those muscles. Men in battle, where fear can be sudden and intense, have a phrase that recognizes the phenomenon of parasympathetic compensation when they suggest that "Only his laundryman knows how frightened he was." Some authors also believe that "voodoo death" may be caused by heart block as a result of parasympathetic compensation due to the intense fear resulting from belief in the voodoo curse.

The **sympathetic trunks** lie on either side of the vertebral column and consist of a series of ganglia connected by neurons. Although connections with the spinal cord start at the **first thoracic vertebra,** the sympathetic chain extends in an anterior direction as far as the first cervical vertebra. At the anterior extremity is the **superior cervical ganglion** (2–68), below that are the **middle cervical ganglion** (3–68) and the **inferior cervical ganglion** (4–68). The nerve bundle running from the superior cervical ganglion is the **internal carotid nerve** (1–68). The ganglion at the level of the first thoracic vertebra is called the **stellate ganglion** (5–68).

The sympathetic neurons that pass through the sympathetic chain without synapsing form the three splanchnic nerves. The **greater splanchnic nerve** (17–68) emerges from the fifth to the ninth thoracic segments; the **lesser splanchnic nerve** (18–68) comes from the tenth and eleventh thoracic segments; and the **least splanchnic nerve** (19–68) comes from the eleventh and twelfth thoracic segments. All three terminate in the **celiac ganglion** (15–68), where they synapse with the postganglionic neurons. The neurons from the lumbar region join with some neurons from the twelfth thoracic region to terminate in the **superior mesenteric ganglion** (22–68) and the **inferior mesenteric ganglion** (25–68), where they synapse with the postganglionic fibers. The points of termination of the postganglionic fibers are shown in Figure 68.

Figure 68

Sympathetic division of the autonomic nervous system.

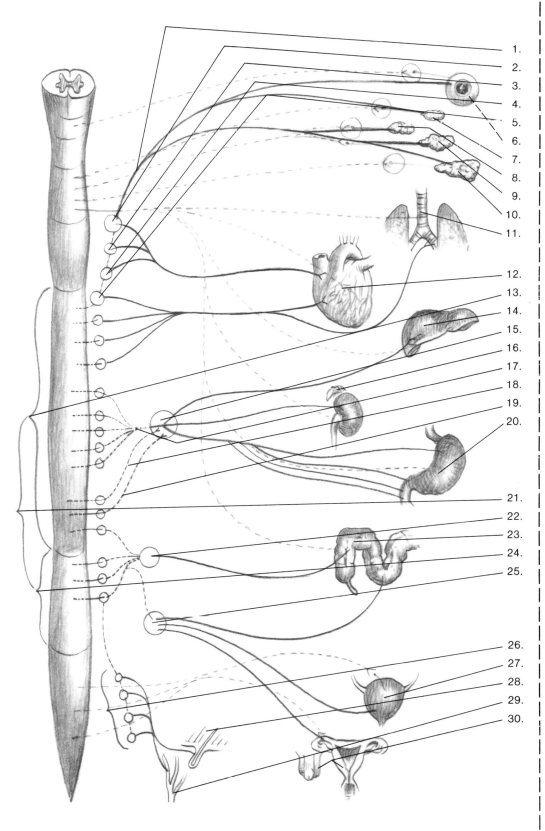

1. | Internal carotid nerve
2. | Superior cervical ganglion
3. | Middle cervical ganglion
4. | Inferior cervical ganglion
5. | Stellate ganglion
6. | Iris and ciliary muscles
7. | Lacrimal gland
8. | Sublingual gland
9. | Submaxillary gland
10. | Parotid gland
11. | Respiratory mechanisms

12. | Heart
13. | Thoracic cord
14. | Liver
15. | Celiac ganglion
16. | Adrenal medulla
17. | Greater splanchnic nerve
18. | Lesser splanchnic nerve
19. | Least splanchnic nerve
20. | Digestive mechanism

21. | Thoracic lumbar outflow
22. | Superior mesenteric ganglion
23. | Colon
24. | Lumbar cord
25. | Inferior mesenteric ganglion

26. | Sacral outflow
27. | Bladder
28. | Hair follicles
29. | Blood vessels
30. | External genitalia

THE PARASYMPATHETIC DIVISION OF THE AUTONOMIC NERVOUS SYSTEM

As is apparent from Figure 68, the **preganglionic neurons** (1–69) of the sympathetic system are generally short, whereas the postganglionic neurons are long. In the parasympathetic division, diagramed in Figure 69, just the opposite is true. The ganglia of the parasympathetic system are located near the organ of innervation. The cranial portion, also called the *cranial outflow,* of the parasympathetic system includes four ganglia. The most anterior is the **ciliary ganglion** (4–69), which is located behind the eye and near the optic nerve [II]. The preganglionic neurons are carried in the **oculomotor nerve [III]** (6–69). The postganglionic neurons innervate the ciliary and iris muscles of the eye.

The **sphenopalatine ganglion** (11–69), also called the *pterygopalatine ganglion,* receives fibers from the **facial nerve [VII]** (14–69). The postganglionic neurons project to the lacrimal gland and to the glands of the nasal and pharyngeal mucosa.

The preganglionic neurons of the **otic ganglion** (10–69) are included in the **glossopharyngeal nerve [IX]** (16–69). The postganglionic neurons from the otic ganglion activate the parotid gland.

The **submandibular ganglion** (13–69) is located near the submaxillary gland. This ganglion receives neurons from the facial nerve [VII] through the **chorda tympani** (15–69), which is one of the branches of the seventh cranial nerve. The postganglionic neurons from the submandibular ganglion innervate portions of the mouth—the sublingual and submandibular salivary glands.

The **vagus nerve [X]** (19–69) contributes the largest portion of the parasympathetic nerves from the cranial outflow. It sends fibers to the entire range of viscera from the lungs to the transverse colon, as Figure 69 shows.

The sacral part of the parasympathetic system, also referred to as the *sacral outflow,* is derived from the second, third, and fourth segments of the sacral cord. These nerves combine to form the **pelvic nerve** (24–69). These preganglionic fibers end in the terminal ganglia located in the pelvic plexus and the myenteric and submucosal plexuses of the colon and rectum. The postganglionic fibers innervate the colon, rectum, urinary bladder, and external genitalia.

Figure 69

Parasympathetic division of the autonomic nervous system.

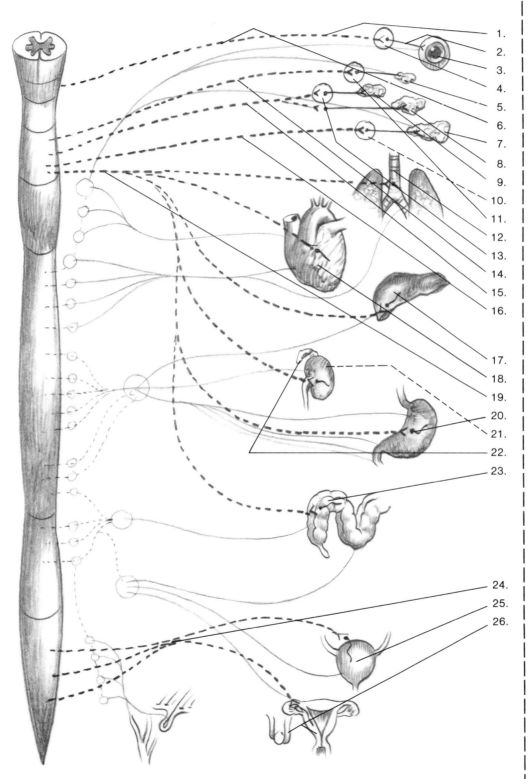

1. Preganglionic neuron
2. Postganglionic neuron
3. Iris and ciliary muscles
4. Ciliary ganglion
5. Lacrimal gland
6. Oculomotor nerve III
7. Parotid gland
8. Submaxillary gland
9. Sublingual gland
10. Otic ganglion
11. *Sphenopalatine ganglion
12. Respiratory mechanisms
13. Submandibular ganglion
14. Facial nerve VII
15. Chorda tympani
16. Glossopharyngeal nerve IX

17. Liver
18. Heart
19. Vagus nerve X
20. Digestive mechanisms
21. Kidney
22. Adrenal medulla
23. Colon

24. Pelvic nerve
25. Bladder
26. External genitalia

PATHWAYS FOR GROSS TACTILE SENSITIVITY

The human ability to perceive contact has a very wide range. At one level the individual is aware only that he has been touched and is unable to localize that contact with any degree of accuracy. At the other level, highly precise judgments can be made about the identity and locus of the stimulation. The amount of precision in tactile sensory judgments is a function of the kind of receptors involved, of the ratio between receptors and neurons, and of the central mechanisms involved.

The neurons that mediate gross, relatively undifferentiated tactile sensibility have their cell bodies in the spinal ganglia of the dorsal root (see Fig. 70). They project heavily myelinated dendrites out to the skin surface, where they terminate in **free nerve endings** (10–70). The axons from these cell bodies enter the dorsal gray matter of the cord and immediately divide into ascending and descending branches, with the ascending branches generally longer than the descending branches. These branches range over as many as eight spinal segments. At varying levels these collaterals synapse with secondary neurons. As a result of the extent of the collaterals, there is considerable overlap among the individual neurons serving the gross tactile sensibility function, thus preventing precise localization.

The cell body of the secondary neuron is located in the posterior horn of the spinal gray matter. Its axon proceeds in a ventromedial direction and crosses to the opposite side through the **anterior commissure** (13–70). These secondary neurons collect in a bundle along the ventrolateral aspect of the cord and proceed in a rostral direction to form the **ventral spinothalamic tract** (14–70). There is evidence that a few of the secondary neurons do not cross to the opposite side of the cord, but ascend in the ispsilateral spinothalamic tract. The fibers in this tract are arranged so that those originating in the more caudal portions of the cord are lateral to those having a more rostral origin. At the level of the medulla some fibers of this tract project to or send collaterals to the lateral part of the brain-stem reticular formation.

The exact course of the ventral spinothalamic tract through the medulla and the pons is not clearly understood as yet. It is believed, however, that it moves somewhat dorsally in the medulla and that by the time it reaches the pons it travels along the dorsal and lateral side of the **medial lemniscus** (6–70) to terminate in the **ventral posterior lateral nucleus of the thalamus** (4–70).

In the thalamus these secondary neurons synapse with tertiary neurons that project to the **postcentral gyrus** (1–70) of the cortex through the **internal capsule** (2–70).

Figure 70

Pathways for gross tactile sensitivity.

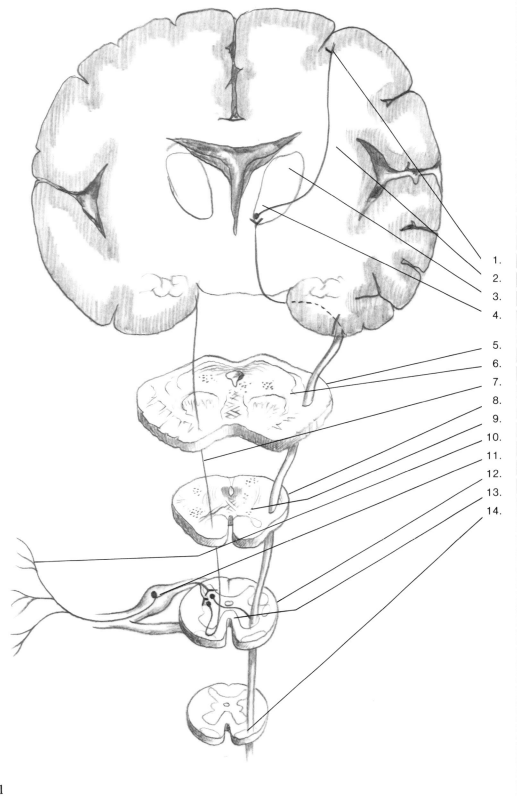

1. Postcentral gyrus
2. Internal capsule
3. *Thalamus
4. Ventral posterior lateral nucleus
5. *Midbrain
6. Medial lemniscus
7. Uncrossed fibers
8. Medulla
9. Reticular substance
10. Free nerve endings
11. Spinal ganglion
12. *Spinal cord
13. Anterior commissure
14. Ventral spinothalamic fasciculus

SECTION 50

PATHWAYS FOR TACTILE DISCRIMINATION AND KINESTHETIC SENSE

The dendrites for the neurons that mediate the highly discriminative tactile sense are heavily myelinated. The cell bodies are in the dorsal root ganglia, and they project out to the skin, ending in a variety of sense organs that permit the fine discrimination of a variety of stimuli. The finest sense of touch is probably mediated in humans by neurons ending in **Meissner's corpuscles** (15–71). The lower the ratio of Meissner's corpuscles to neurons, the finer the discrimination. The greatest concentration of corpuscles is probably in the tip of the index finger, where frequently there is only one corpuscle to a single neuron. The vibratory sense is most likely related to the **Pacinian corpuscles** (16–71), which are located in the subcutaneous connective tissue.

The neurons mediating the kinesthetic and proprioceptive senses, which permit the individual to be aware of the location or movement of parts of the body, also have their cell bodies in the dorsal root ganglia and are also heavily myelinated. The dendrites terminate in **muscle spindles** (9–71) and **unencapsulated joint receptors.**

Neurons for both the tactile sense and the kinesthetic sense send axons into the spinal cord medial to the dorsal horn cells (Fig. 71). At that point they bifurcate and produce ascending and descending branches, which travel up and down the cord in the posterior white columns. As the fibers from the lower levels of the cord travel upward, they are shifted in a medial and dorsal direction by the new fibers of the same type coming in from the higher levels. The result is that the fibers from the lower extremities are located in the medial posterior part of the cord, while those of the higher levels of the body and the upper extremities are in a posterolateral position.

At the level of the upper thoracic and cervical sections of the cord the posterior white column is divided in two by the posterior intermediate sulcus, producing two distinct fasciculi—the medially located **fasciculus gracilis** (7–71) and the more lateral **fasciculus cuneatus** (8–71). The fibers in these two fasciculi travel in a rostral direction to the medulla, where the fasciculus gracilis terminates in the **nucleus gracilis** (11–71) and the fasciculus cuneatus terminates in the **nucleus cuneatus** (10–71).

The secondary neurons in this system have their cell bodies in the nuclei gracilis and cuneatus. Their axons proceed in a ventromedial direction as the **internal arcuate fibers** (12–71) of the medulla. They cross the midline and ascend in the well-defined bundle of fibers known as the **medial lemniscus** (5–71). The medial lemniscus travels through the pons and midbrain to terminate in the **ventral posterior lateral nucleus of the thalamus** (4–71). **Tertiary neurons** (2–71) extend from the thalamus to the **postcentral gyrus** (1–71) of the cortex by way of the **posterior limb of the internal capsule** (3–71).

152

Figure 71

Pathways for tactile discrimination and the kinesthetic sense.

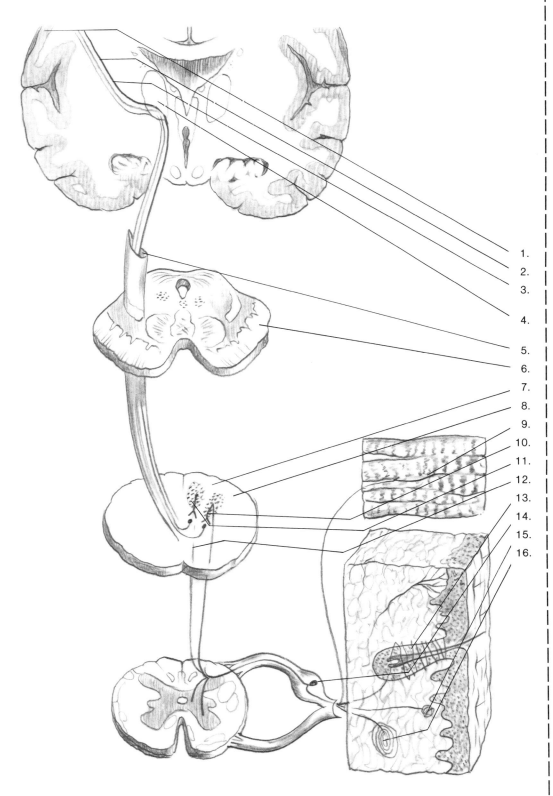

1.   Postcentral gyrus
2.   Tertiary neurons
3.   Posterior limb of internal capsule
4.   Ventral posterior lateral nucleus of thalamus
5.   Medial lemniscus
6.   *Midbrain
7.   *Fasciculus gracilis
8.   *Fasciculus cuneatus
9.   Muscle spindle
10.   Nucleus cuneatus
11.   Nucleus gracilis
12.   Internal arcuate fibers
13.   Nerve endings on hair cell
14.   Dorsal root ganglion
15.   Meissner's corpuscle
16.   Pacinian corpuscle

PATHWAYS FOR PAIN AND TEMPERATURE SENSE

The neurons that convey the sense of pain from the viscera and the periphery of the body have their cell bodies in the dorsal root ganglia (see Fig. 72). They are unmyelinated and end in the free nerve endings throughout the body. These fibers are of very small diameter, and they conduct impulses relatively slowly compared with the large fibers serving touch and proprioception.

The fibers of pain enter the cord and bifurcate in the area of the tip of the dorsal horn. The branches generally do not travel more than one segment in either direction before they synapse with small, lightly myelinated fibers that cross the midline in the **anterior commissure** (15–72) and proceed to the lateral portion of the opposite side of the cord; here they collect in an ascending bundle to form the **lateral spinothalamic fasciculus** (8–72). Fibers from the lower portion of the cord are displaced dorsally and laterally by those entering the tract at a higher level. The result is a topographical representation within the lateral spinothalamic tract with the more rostral portions of the body in the ventral portion and the more caudal in the dorsal portion.

On entering the cord the pain fibers also send collaterals to the **dorsolateral fasciculus** (13–72), or *tract of Lissauer,* on the same side. The fibers of the dorsolateral fasciculus are relatively short, but they are distributed throughout the spinal cord. Thus, some representation of the sensation of pain is ipsilateral as well as contralateral. The fibers of this tract also make intrasegmental and intersegmental connections with motor nerves in the ventral horn cells and provide for reflex action to painful stimuli.

The neurons serving the temperature sense follow a course similar to that of the neurons for pain (see Fig. 72). In the skin they terminate in the **Krause end bulbs** (11–72) and in the **brushes of Ruffini** (12–72) which are believed to be associated with temperature perception. Within the lateral spinothalamic tract, they are intermingled with the pain fibers, but more temperature fibers tend to be concentrated in the posterior portion of the tract while the pain fibers are more anterior.

The spinothalamic tract ascends through the medulla, the pons, and the midbrain and sends collaterals to the **reticular substance** (10–72) in each of those areas. Ultimately the tract reaches the thalamus and terminates in the **ventral posterior lateral nucleus** (4–72). Tertiary neurons run from there to the **postcentral gyrus** (1–72) in the parietal lobe.

Since the pain fibers ascend in the neat bundle of the lateral spinothalamic fasciculus, it is possible to relieve intractable pain with an operation called a chordotomy. In this operation a lesion is made in the lateral spinothalamic fasciculus that produces anesthesia and loss of temperature sense on the side of the body opposite the lesion. In some cases pain may return because it is transmitted by the short fibers of the dorsolateral tract. Visceral pain is represented bilaterally and can therefore only be relieved by lesioning the lateral spinothalamic fasciculus on both sides.

The lateral spinothalamic fasciculus can also be lesioned in the medulla or in the midbrain, where it is reasonably accessible because of its closeness to the surface. This operation can be used to relieve intractable pain in the more rostral portions of the body.

Figure 72

Pathways for pain and temperature sense.

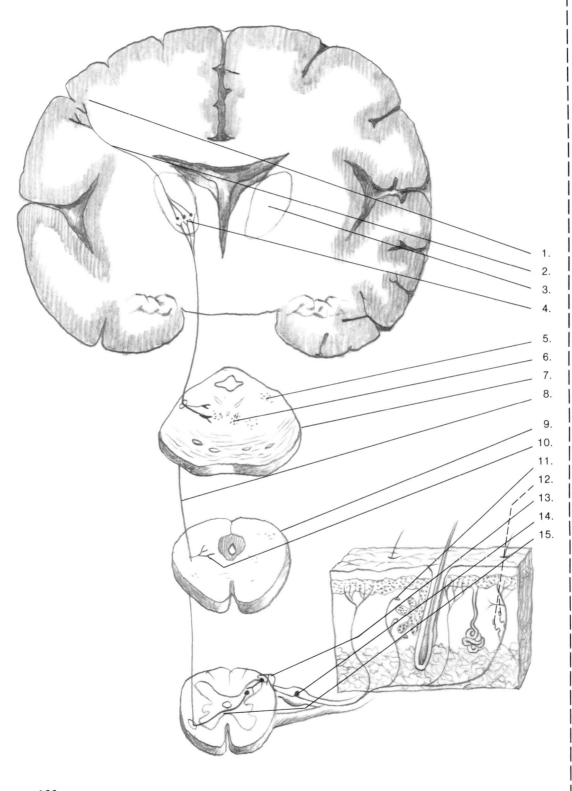

1.   Postcentral gyrus
2.   Internal capsule
3.   *Thalamus
4.   Ventral posterior lateral nucleus
5.   Reticular substance
6.   Oral pontine reticular nucleus
7.   Pons
8.   Lateral spinothalamic fasciculus
9.   Medulla
10.   *Reticular substance
11.   Krause end bulbs
12.   Brushes of Ruffini
13.   *Dorsolateral fasciculus
14.   Spinal ganglion
15.   Anterior commissure

PATHWAYS FOR VISION

The eye and the associated pathways are actually an extension of the central nervous system. (See Fig. 73.) Embryonically, the visual apparatus is derived from an evagination of the diencephalon that migrates to the periphery. Thus, the components of the visual system are more closely related to central-nervous-system tissue than to peripheral receptor mechanisms.

The photosensitive visual receptors are the rods and the cones located in the retina of the eye. There are three major cell layers in the **retina** (8–73), including the **ganglion cells** (6–73), the **bipolar cells** (4–73), and the photosensitive cells that are the **rods** (2–73) and the **cones** (1–73). It is interesting that the light-sensitive cells are under the others. The light rays must go through two cell layers before activating the receptor cells.

In addition to the three primary cell types, the retina contains **horizontal cells** (3–73) and **amacrine cells** (5–73), which contribute to the integration of impulses generated in the light-sensitive cells. **Pigmented cells** line the retina and give it a dark color.

Although estimates vary, there are said to be approximately 6.5 million cones and 125 million rods. In the optic tract, however, there are about 1.25 million fibers. It has been said that 38 percent of all of the fibers entering or leaving the central nervous system are in the optic nerve [II].

The cones contribute to the perception of color and fine detail and are concentrated in the **macula lutea** (7–73), a small, yellowish, circular area at the back of the retina in a direct line with the visual axis. In the center of the macula lutea is an area composed of slender cones that are closely packed together. This is the **fovea centralis,** the area of sharpest vision and most discriminating color perception.

The rods that are responsible for the perception of dim light are totally absent in the fovea but become more prevalent in the periphery of the retina. Both the rods and the cones are connected to the ganglion cells through bipolar cells. In the area where vision is most acute, a single cone is connected to a single bipolar cell. In the rest of the retina, however, varying numbers of the light-sensitive cells may send impulses to a single bipolar cell. In some cases, one bipolar cell may synapse with both rods and cones.

The ganglion cells that make up the fibers of the optic tract may synapse with a single bipolar cell in the area of central vision or may serve a number of bipolar cells in the periphery.

The fibers of the **optic nerve [II]** (9–73) become myelinated as they leave the eyeball and proceed through the optic foramen. The nerves from the two eyes join to form the **optic chiasm** (10–73). They emerge from the chiasm as the **optic tract** (11–73), which terminates primarily in the cells of the **lateral geniculate body** (12–73).

Within the optic chiasm a partial decussation of the fibers of the two nerves takes place. Those fibers originating in the temporal portion of each retina continue through the chiasm without crossing, while those fibers coming from the nasal portion of the retina of the two eyes cross to the opposite side. As a result, the optic tract is composed of fibers from both eyes.

About 80 percent of the fibers of the optic tract terminate in the geniculate body, while the other 20 percent project to the **superior colliculi** (13–73) and the **pretectal area.** The latter contribute the afferent input to the various visual reflexes.

The final relay of the visual pathway is the **geniculo-calcarine tract** (14–73), which runs from the lateral geniculate body through the posterior limb of the internal capsule and terminates in the visual cortex of the occipital lobe. The visual cortex lies on the medial surface on both sides of the **calcarine sulcus** (15–73).

There is a precise, point-to-point relationship between the retinal areas and the lateral geniculate body and between the lateral geniculate and the striate cortex. As a result, damage to any particular portion of the visual projection system produces predictable and specific losses of visual ability.

Figure 73

Pathways for vision.

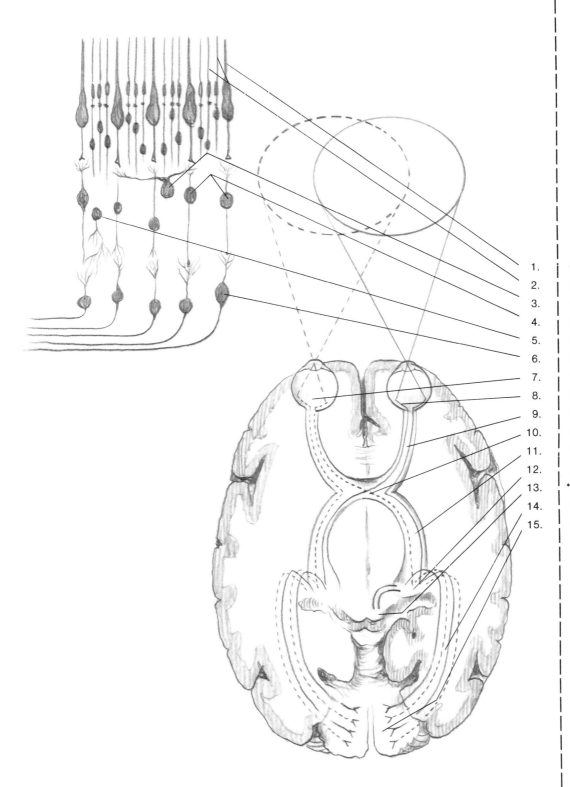

1.   Cone
2.   Rod
3.   Horizontal cell
4.   Bipolar cell
5.   Amacrine cell
6.   Ganglion cell
7.   Macula lutea
8.   Retina
9.   Optic nerve II
10.   Optic chiasm
11.   Optic tract
12.   Lateral geniculate body
13.   *Superior colliculus
14.   Geniculocalcarine tract
15.   Calcarine sulcus

PATHWAYS FOR AUDITION

The cell bodies of the primary neurons of audition are located in the **spiral ganglion** (18–74) near the inner ear. The dendrites from those bipolar ganglion cells end in the hair cells of the **spiral organ of Corti** (19–74). These hair cells convert sound waves into neural activity. Different portions of the spiral organ of Corti are sensitive to different frequencies of sound. The higher frequencies are appreciated in basal coils of the spiral organ, while the low tones activate the apical portion.

The axons of the cells of the spiral ganglia constitute the acoustic nerve [VIII]. As the fibers enter the brain stem on the lateral side of the pons, they bifurcate; the ascending branches go to the **dorsal cochlear nucleus** (12–74), while the shorter, descending branches go to the **ventral cochlear nucleus** (14–74). There is a localization according to tone in the various levels of the auditory projections, so that particular portions of the spiral organ of Corti project to particular parts of the auditory system.

The pathways from the cochlear nuclei to the cortex are complex and not yet completely worked out. However, the following description provides a general schema. (See also Fig. 74.)

Secondary fibers from the dorsal cochlear nucleus project across the midline to terminate in the **superior olivary nucleus** (11–74) or they continue and ascend in the **lateral lemniscus** (9–74) of the opposite side. Some cross in the **trapezoid body** (13–74) and may terminate in its nucleus. Some of the secondary fibers from the dorsal cochlear nucleus do not cross the midline but project directly to the superior olivary nucleus on the same side and may join the ipsilateral lateral lemniscus.

The ventral cochlear nucleus sends second-order neurons to the superior olivary nucleus on the same side or through the trapezoid body to the superior olivary nucleus on the opposite side. Some of them terminate in the **nuclei of the trapezoid body** (21–74). Others join the lateral lemniscus on the same and opposite sides. Both cochlear nuclei project fibers to the reticular formation.

The lateral lemniscus, which receives fibers from both cochlear nuclei on both sides, proceeds in a rostral direction and terminates in the **inferior colliculus** (7–74) and the **medial geniculate body** (3–74). Masses of gray matter are located within the lateral lemnisci and are known as the **nuclei of the lateral lemniscus** (10–74). The **inferior brachium** (5–74) contains connecting fibers from the inferior colliculi to the medial geniculate bodies. The two inferior colliculi are connected by the **commissure of the inferior colliculus** (6–74).

The medial geniculate body receives fibers from the cochlear nuclei of both sides as well as from both superior olivary nuclei. It receives many fibers from the inferior colliculus on the same side and a few from the opposite side. The two medial geniculate bodies are connected by the **commissure of Gudden** (4–74). Axons from the medial geniculate body project through the posterior limb of the internal capsule to the primary auditory area in the temporal lobe, the transverse temporal gyrus.

Figure 74

Pathways for audition.

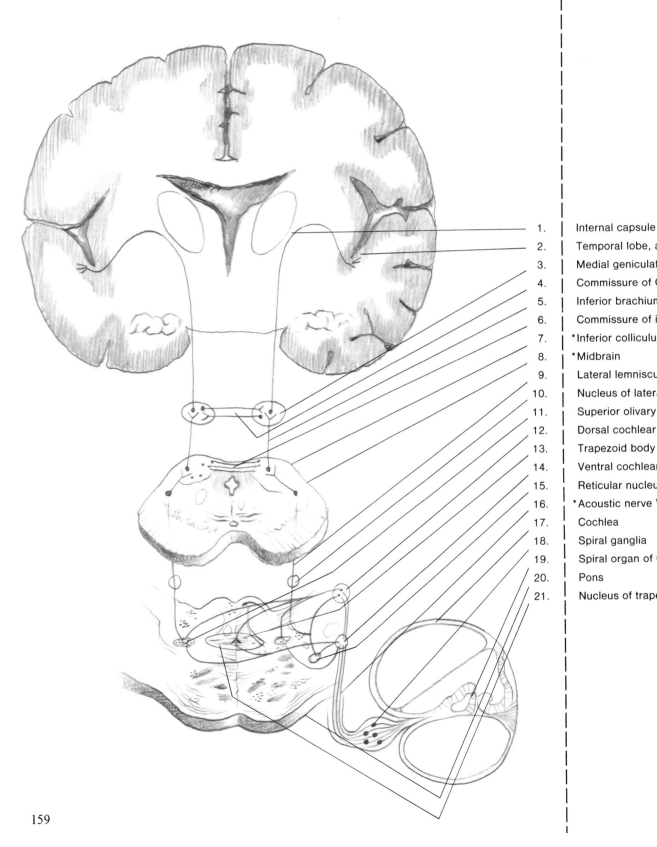

1. Internal capsule
2. Temporal lobe, auditory area
3. Medial geniculate body
4. Commissure of Gudden
5. Inferior brachium
6. Commissure of inferior colliculi
7. *Inferior colliculus
8. *Midbrain
9. Lateral lemniscus
10. Nucleus of lateral lemniscus
11. Superior olivary nucleus
12. Dorsal cochlear nucleus
13. Trapezoid body
14. Ventral cochlear nucleus
15. Reticular nucleus
16. *Acoustic nerve VIII
17. Cochlea
18. Spiral ganglia
19. Spiral organ of Corti
20. Pons
21. Nucleus of trapezoid body

PATHWAYS FOR THE VESTIBULAR SYSTEM

The primary neurons for the vestibular mechanism (see Fig. 75) are located in the internal auditory meatus in the **vestibular ganglion** (10–75), or *ganglion of Scarpa*. The dendrites of these bipolar neurons terminate in the primary receptor mechanisms of the semicircular canals. The three canals are oriented at right angles to one another and are fixed in the three planes in space. At the base of each canal is an enlarged cavity, the **ampulla** (17–75), in which the vestibular receptors are located. These receptors are called **cristae** (singular, **crista**). The hairs of the cells of the cristae are covered with a gelatinous mass called the **cupula.** The semicircular canals are filled with an endolymphatic fluid that has inertia when still. Any movement of the head produces a flow in the fluid, which acts on the cristae, producing the adequate stimulus for the vestibular neurons. The principal source of stimulation for the receptors in the ampullae is rotatory and angular acceleration.

The second portion of the vestibular apparatus consists of the **utricle** (19–75) and the **saccule** (18–75). The ends of the semicircular canals open into the utricle, which is connected to the saccule. The **maculae** are the sensory receptors contained within these two cavities. The hair cells of the maculae are also covered with a gelatinous mass in which is embedded small crystals of calcium carbonate called *otoliths.* The maculae are activated by linear movement or changes in the position of the head.

The axons of the cells from the vestibular ganglion form the **vestibular nerve [VIII]** (11–75), which enters the brain stem just medial to the acoustic nerve [VIII]. After entering the brain the fibers of the vestibular nerve bifurcate into short, ascending branches and relatively long, descending branches. Some of the ascending branches proceed directly to the cerebellum, terminating primarily in the cortex of the **nodulus** (28–12), **uvula** (35–12), and **flocculus** (26–12). Most of the vestibular fibers, however, terminate in one of the four vestibular nuclei: the **medial vestibular nucleus** (5–75), also called *Schwalbe's nucleus,* the **lateral vestibular nucleus** (8–75), also called *Deiters' nucleus,* the **superior vestibular nucleus** (6–75), also referred to as the *Bechterew's nucleus,* and the **spinal vestibular nucleus** (9–75).

The medial vestibular nucleus projects a large number of fibers to the **medial longitudinal fasciculus** (13–75) on both sides of the brain. These fibers bifurcate, producing both ascending and descending branches. Neurons from the medial nucleus also connect with nuclei of

the reticular formation in various parts of the medulla and the pons.

The lateral vestibular nucleus sends fibers into the reticular formation. Fibers from this nucleus also form the direct, uncrossed **vestibulospinal tract** (14–75). Collaterals from the lateral nucleus ascend through the **inferior cerebellar peduncle** (24–12) into the cerebellum to terminate in the **vermis** (9–82).

The superior vestibular nucleus contributes fibers to the medial longitudinal fasciculus and sends fibers to the cerebellum that terminate in the **vermis** (9–82) and the **flocculonodular lobe** (26–13).

The **spinal vestibular nucleus** (9–75) has connections with the reticular nuclei of the medulla and pons and sends fibers to the medial longitudinal fasciculus on both sides, and to the *accessory nerve nucleus* which contributes to the control of head and neck muscles.

Fibers from the superior, medial, and lateral nuclei ascend to connect with the nuclei of the three cranial nerves that control eye movements: the **abducens [VI]** (4–75), the **trochlear [IV]** (3–75), and the **oculomotor [III]** (2–75) nerves.

Figure 75

Pathways for the vestibular system.

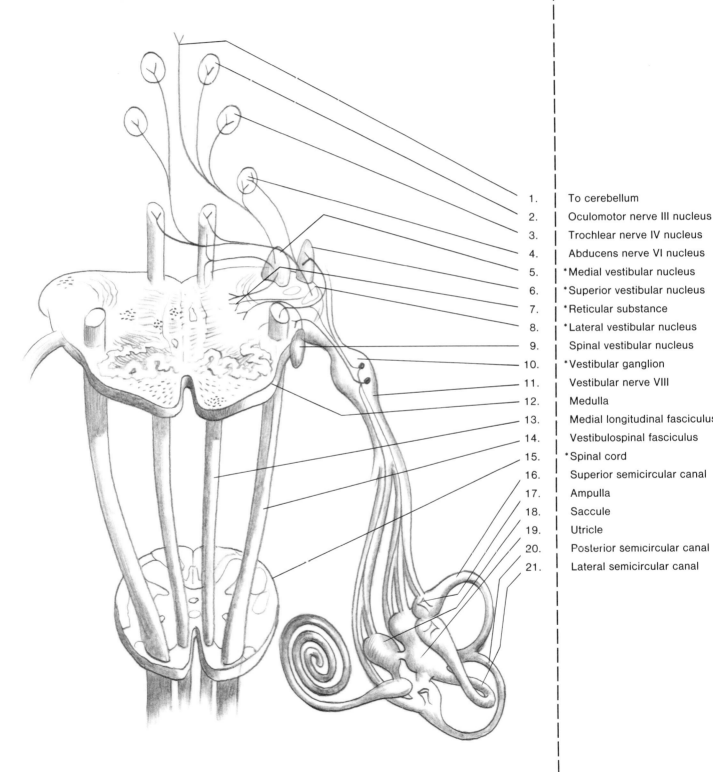

1. To cerebellum
2. Oculomotor nerve III nucleus
3. Trochlear nerve IV nucleus
4. Abducens nerve VI nucleus
5. *Medial vestibular nucleus
6. *Superior vestibular nucleus
7. *Reticular substance
8. *Lateral vestibular nucleus
9. Spinal vestibular nucleus
10. *Vestibular ganglion
11. Vestibular nerve VIII
12. Medulla
13. Medial longitudinal fasciculus
14. Vestibulospinal fasciculus
15. *Spinal cord
16. Superior semicircular canal
17. Ampulla
18. Saccule
19. Utricle
20. Posterior semicircular canal
21. Lateral semicircular canal

PATHWAYS FOR THE SENSE OF TASTE

The transducers for the special sense of taste (see Fig. 76) are the taste buds, which are located in the mucous membrane of the tongue, palate, pharynx, and larynx. Generally what the individual perceives as the sense of taste is, in fact, a combination of both olfactory and gustatory stimulation. It is well known that to discriminate between the taste of an apple and the taste of an onion is difficult if olfaction is blocked by holding the nose.

The taste buds are located in **papillae** (12–76), which are ridges of tissue that can be readily seen on the edges and the superior surface of the tongue. The taste buds contain a number of elongated cells that project a delicate hairlike process into an opening called a *gustatory pore.* Only these hairlike processes, the **microvilli,** are in contact with the chemical stimuli for taste.

There are four basic taste qualities: sweet, sour, bitter, and salty. Any taste bud may be sensitive to one or more of these four qualities, but there is some tendency for concentration of the four different kinds of taste in different portions of the tongue. Sweetness is appreciated more readily at the tip of the tongue, sour at the sides, and bitter at the back. The appreciation of the sensation of saltiness is more widely distributed.

The taste buds of the anterior two-thirds of the tongue are supplied by the afferent neurons of the **facial nerve [VII]** (9–76). The posterior two-thirds of the tongue are supplied by afferents from the **glossopharyngeal nerve [IX]** (10–76). Afferent fibers from the **vagus nerve [X]** (11–76) supply the taste buds of the pharynx and larynx.

The axons of all of the nerves relating to taste project to the **nucleus solitarius** (8–76). The course of the secondary fibers for taste is open to some debate. However, there is evidence that they form an ascending gustatory tract within the **medial lemniscus** (6–76) and terminate in the **posteromedial ventral nucleus** (4–76) of the thalamus. Tertiary fibers project from there to a portion of the cortex near the tactile-projection area of the tongue and to adjacent areas, including the insula.

Figure 76

Pathways for the sense of taste.

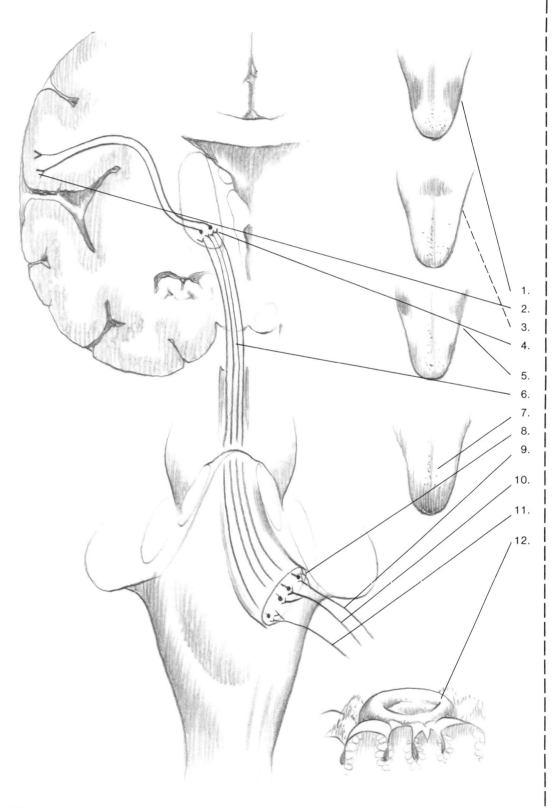

1. Salt
2. Postcentral gyrus
3. Bitter
4. Posteromedial ventral nucleus of thalamus
5. Sour
6. Medial lemniscus
7. Sweet
8. Nucleus solitarius
9. Facial nerve VII, anterior 2/3 of tongue
10. Glossopharyngeal nerve IX, posterior 2/3 of tongue
11. Afferent vagus nerve X, taste buds, larynx, and pharynx
12. Papillae

PATHWAYS FOR THE OLFACTORY SENSE

Olfaction is a chemical sense; thus stimuli must be in solution before they can activate the receptors for smell. This is accomplished, in part, by the fact that the olfactory epithelium is constantly bathed in a solution. It has been suggested repeatedly that humans are microsomic with the implication that they are insensitive to odors. While it is true that some animals have a better developed sense of smell, the human olfactory apparatus is remarkably sensitive. The receptors will respond to one part of vanillin in 10 million parts of air. The odor of skunk (mercaptans) will activate the olfactory receptors if there are 123 millionths of a milligram in a liter of air. It has been estimated that as few as forty molecules of mercaptans are sufficient for detection in humans, and it may be that a single molecule will cause a single receptor to fire.

The **olfactory epithelium** (16–77) is located in a small area that extends over the lateral and medial walls and the top of the nasal meatus. The olfactory nerve [I] is the shortest of the cranial nerves. It runs just a few millimeters from the olfactory mucous membrane to the olfactory bulb.

The primary neurons of olfaction are bipolar and contain a small amount of cytoplasm and a large nucleus, with the result that the cell is expanded in the area of the nucleus. Each cell has a slender peripheral process that extends out to the free surface of the mucous membrane. These processes terminate in a tuft of very fine **olfactory hairs** (17–77). Centrally, the olfactory cells taper into long, thin, unmyelinated threads, the *olfactory fila,* which form into bundles and pass through the **cribriform plate** (15–77) of the ethmoid bone. Collectively, these fibers are referred to as the **olfactory nerve [I]** (14–77). These cells terminate in the **olfactory bulb** (13–77).

Within the olfactory bulb the primary cells synapse with two types of cells, the relatively large mitral cells and the tufted cells to form **glomeruli** (3–77). The axons of the mitral cells and the tufted cells constitute the secondary olfactory fibers and enter the **olfactory tract** (10–77).

A third set of neurons, which are between the mitral and the tufted cells in size, synapse with those cells and form the **anterior olfactory nucleus** (9–77). The axons of the cells from the anterior olfactory nucleus proceed centrally in the olfactory tract and go through the anterior portion of the **anterior commissure** (4–77) to synapse with the cells of the anterior olfactory nucleus on the opposite side.

The olfactory tract travels centrally and divides into the **lateral olfactory stria** (8–77) and the **medial olfactory stria** (7–77). The fibers of the lateral olfactory stria terminate in the **prepyriform cortex** (12–77) and in the **anterior nucleus of the amygdala** (11–77). This is the primary olfactory cortex. Olfaction is the only sensory system that sends fibers directly to the cortex without a relay in the thalamus. Fibers from the medial olfactory stria terminate in the **subcallosal gyrus** (6–77) and the **parolfactory area** (5–77).

It was once believed that the entire rhinencephalon (see Sect. 40) was involved in the sense of smell. However, it is now generally agreed that the direct central olfactory connections are restricted to those indicated in Figure 77.

Figure 77

Pathways for the olfactory sense.

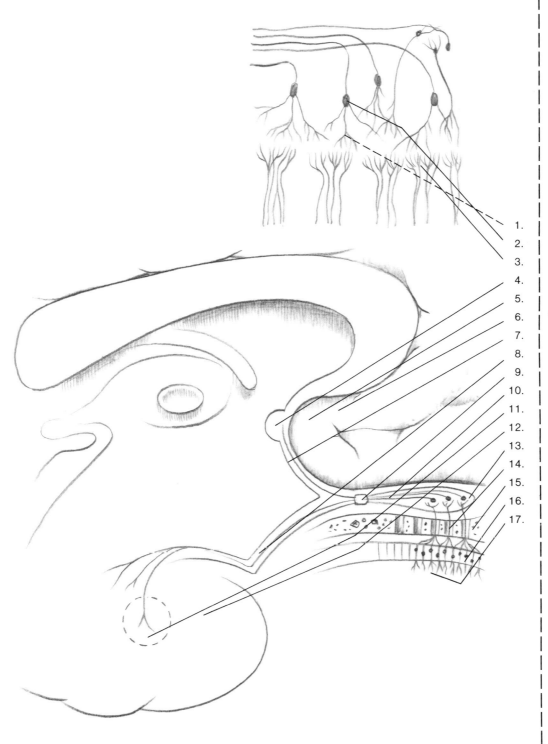

1. Tufted cell
2. Mitral cell
3. Glomerulus
4. Anterior commissure
5. Parolfactory area
6. *Subcallosal gyrus
7. Medial olfactory stria
8. Lateral olfactory stria
9. Anterior olfactory nucleus
10. Olfactory tract
11. Amygdala, anterior nuclei
12. Prepyriform cortex
13. Olfactory bulb
14. Olfactory nerve I
15. Cribriform plate
16. Olfactory epithelium
17. Olfactory hairs

THE PYRAMIDAL SYSTEM

The large motor system concerned with voluntary muscular control has its origin primarily in the precentral gyrus of the cerebral cortex. Motor control for the various parts of the body are arranged in an orderly fashion according to the homunculus schematically presented on the cortical representation in Figure 78. It can be seen that the homunculus is situated head down with its leg hanging over the longitudinal fissure. The amount of cortex devoted to the control of a portion of the body is a function of how fine that control needs to be. Thus, there is relatively little of the precentral gyrus available for control of the hip and the trunk, but the muscles of vocalization and those for finger control have a considerable representation. The area that controls the ankle, foot, and toes is in the medial aspect of the cortex inside the longitudinal fissure.

The pyramidal tract is the largest and the longest descending fiber system in the human central nervous system. Each tract contains nearly a million neurons about three-quarters of which are myelinated. In man the fibers may be over two feet long. (In the whale they may be as much as 30 feet long.)

Most of the fibers of the pyramidal tract originate in the giant cells of Betz, which are located in the fifth layer in the cortex of the **precentral gyrus** (2–78). Recent evidence indicates that a significant portion of the fibers that make up the pyramidal system come from other portions of the cortex, including the prefrontal, parietal, temporal, and occipital areas.

From their relatively diffuse origin, the fibers collect in the **corona radiata** (3–78) and descend through the **genu** (4–78) and the anterior two-thirds of the posterior portion of the **internal capsule** (5–78). Those fibers in the genu are referred to as the **geniculate fibers** (4–78), while the rest constitute the **cerebrospinal fibers** (5–78). The fibers twist as they descend so that those that control leg movements are located in the most posterior portion of the internal capsule, while those that control facial movements pass through the genu. As the tract proceeds in a caudal direction, the geniculate fibers branch off, cross the midline, and make connections with the cranial nerve nuclei, which ultimately innervate the various muscles of the head area.

The rest of the fibers continue to descend and at the level of the medulla come near the surface as the **pyramids** (14–78). At the junction of the medulla and the cord, about 80 percent of the fibers, those most medially located, cross to the opposite side through the great **pyramidal decussation** (15–78). They continue in a caudal direction through the **lateral corticospinal tract** (18–78). The more laterally placed fibers continue into the cord without crossing in the **anterior corticospinal tract** (16–78). Most but not all of those fibers do cross the cord in the anterior commissure at the point at which they make connections with the motor neurons that go to the periphery. Finally, there is a third tract, which is quite small, that descends in the lateral corticospinal tract without crossing to the opposite side first.

The fibers from the various corticospinal tracts terminate in the gray matter of the spinal cord. Some of the fibers connect directly with the motor neurons that proceed to the periphery and innervate the muscles. Others connect with internuncial neurons that, in turn, synapse with the motor fibers. Since almost all of the fibers of the three corticospinal tracts cross to the opposite side of the cord either at the decussation or below in the cord itself, any lesion in the primary motor system will result in a dysfunction of the muscles on the opposite side of the body.

Approximately 55 percent of the pyramidal fibers terminate in the cervical region of the cord, and 20 percent end in the thoracic section. Only 25 percent proceed to the lumbar and sacral segments. Thus, there is much better innervation and consequently better motor control in the upper body and extremities.

There is evidence that the larger corticospinal tract fibers are primarily concerned with the fine voluntary control of more distal muscles in the hands and feet. The smaller fibers connect with the motor neurons that innervate the more proximal musculature of the trunk.

Figure 78

The pyramidal system.

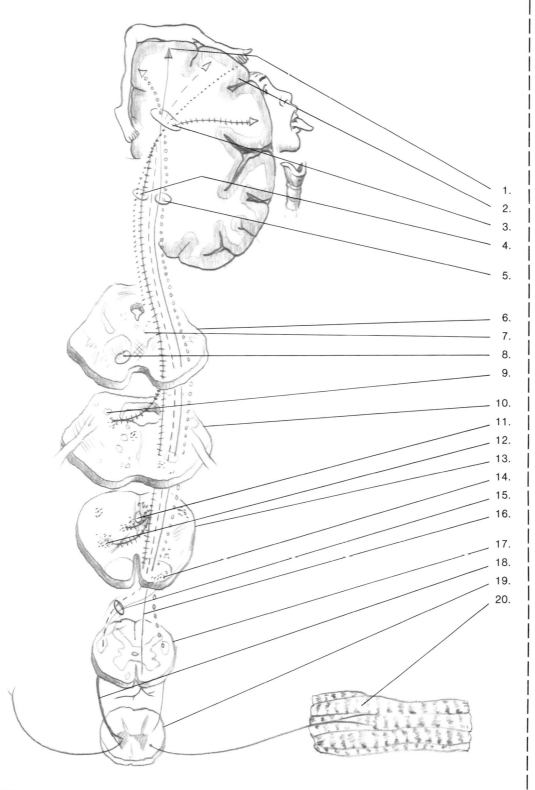

1. *Giant cells of Betz
2. Precentral gyrus
3. Corona radiata
4. Genu of internal capsule, geniculate fibers
5. Posterior limb of internal capsule, cerebrospinal fibers
6. *Midbrain
7. Oculomotor nerve III
8. Red nucleus
9. Motor nucleus of trigeminal nerve V
10. Pons
11. Hypoglossal nerve XII nucleus
12. Nucleus ambiguus
13. Medulla
14. Pyramid
15. Pyramidal decussation
16. Anterior corticospinal fasciculus
17. Cervical cord
18. Lateral corticospinal fasciculus
19. *Spinal cord
20. Muscle

THE EXTRAPYRAMIDAL SYSTEM

While the pyramidal system described in the previous section is in primary control of voluntary movement in humans, it can only function adequately if it operates against a background of automatic postural adjustments, muscle tone, and a servomechanism type of feedback system. The nonvoluntary motor mechanisms are provided, in part, by the extrapyramidal system.

Literally, the extrapyramidal system includes all descending fibers from the cortex that are not included in the pyramidal system. However, custom has restricted the definition of extrapyramidal to the interactions of those subcortical nuclear masses that function to complement the action of the pyramidal system.

Although the entire set of interactions within the extrapyramidal system are too complex to be illustrated on a single diagram, the basic mechanisms are shown in Figure 79. They will be briefly described here.

Motor fibers originate in several different portions of the cortex, particularly in the premotor area, and send fibers to the **globus pallidus** (6–79), the **putamen** (4–79), the **red nucleus** (12–79), and the **reticular substance** (11, 16–79). The globus pallidus also receives fibers from the **caudate nucleus** (2–79) by way of the putamen, and from the putamen itself.

Two major fiber bundles leave the globus pallidus and enter the **internal capsule** (5–79). These are the **fasciculus lenticularis** (7–79) and the **ansa lenticularis** (10–79). Both fiber bundles turn in a dorsal direction to enter the ventral thalamus as the **fasciculus thalamicus** (8–79). Fibers from the fasciculus lenticularis and the ansa lenticularis also turn ventrally and send branches to the **subthalamic nuclei** (9–79) and the area of the red nucleus. Fibers also run from the globus pallidus to the **substantia nigra** (14–79), the **dentate nucleus** (15–79) of the cerebellum, and the **reticular substance** (11,16–79).

Feedback circuits from most of the lower nuclei, including the substantia nigra, the dentate nucleus, the red nucleus, and the reticular system, return to the globus pallidus and to the thalamus. There is also input from the thalamus back to the premotor cortex.

In some diseases of the nervous system, portions of the extrapyramidal system degenerate with a resultant variety of motor dysfunctions. Parkinson's disease, or paralysis agitans, for example, results from a degeneration of portions of the basal ganglia and the substantia nigra. Symptoms of this disorder are easily recognized. They include muscular rigidity, tremor, and a masklike facial expression. The patient stands with head and shoulders stooped and walks with a peculiar shuffling gate leaning forward. It is believed that the extrapyramidal lesions produce a release of the suppression of the output from the globus pallidus, which results in the excessive muscle tone and tremor. A surgical technique has been developed in which lesions are made in the globus pallidus. This therapeutic measure, called *pallidectomy,* reduces the tremor on the opposite side of the body.

Figure 79

Schematic of some pathways involved in the extrapyramidal system.

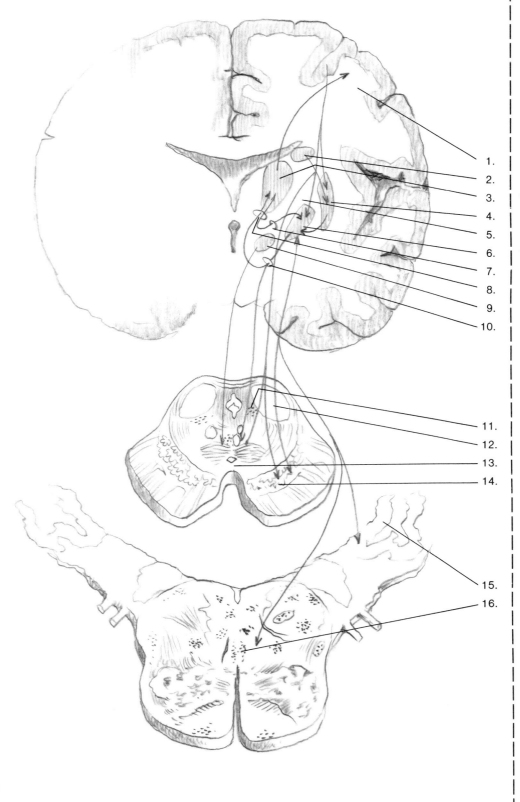

1. Premotor cortex
2. Caudate nucleus
3. *Thalamus
4. Putamen
5. Internal capsule
6. *Globus pallidus
7. Fasciculus lenticularis
8. Fasciculus thalamicus
9. Subthalamic nucleus
10. Ansa lenticularis

11. Midbrain reticular substance
12. Red nucleus
13. Prerubral field
14. Substantia nigra

15. Dentate nucleus
16. Reticular substance

THE CEREBELLUM: PEDUNCLES AND CONNECTIONS

The cerebellum is a critical portion of the extrapyramidal system. It is involved in some measure in all reflexive postural adjustments, and it functions to coordinate all intentional movements, providing both facilitation and inhibition to smooth out all motor activity. It receives afferent fibers from all of the sensory mechanisms and projects efferent fibers to all segments of the nervous system. There is recent evidence that highly organized, well-coordinated behavior such as predatory aggression can be elicited by direct electrical stimulation of portions of the **fastigial nucleus** (11–81), (4–82). The cerebellum is no longer considered to be simply the "head ganglion" for the proprioceptive and vestibular senses.

Sections 59, 60, and 61 will be devoted to the anatomy of the cerebellum. Figure 80 presents the relationship between the cerebellum and the rest of the brain, including the peduncles. Figure 81 shows a schematic of the connections discussed here. Separate sections are devoted to the cerebellar nuclei and the cellular structure of the cortex.

Three major fiber bundles called *peduncles* connect the cerebellum with the rest of the brain. Each is named according to its relative position, and each has a second name given by early anatomists. Although the names based on position will be used throughout this book, you should also learn the older nomenclature because it is frequently used in other textbooks. The three large projection bundles are the **inferior cerebellar peduncle** (8–80), or *restiform body,* the **middle cerebellar peduncle** (7–80), or *brachium pontis,* and the **superior cerebellar peduncle** (4–80), which has also been given the name *brachium conjunctivum.*

The inferior cerebellar peduncle, or *restiform body,* actually enters the cerebellum between the superior and the middle peduncles. The fibers come from the spinal cord and form a portion of the lateral wall of the fourth ventricle. They proceed in a rostral direction, turn laterally, and then turn sharply in a dorsal direction to enter the cerebellum. The inferior peduncle is composed of the following fiber tracts from the spinal cord: (1) The **dorsal spinocerebellar tract** (19–81), which terminates primarily in the **vermis** (9–82), is a proprioceptive pathway that comes from the dorsal nucleus of the same side of the cord. (2) The **olivocerebellar fasciculus** (18–81) terminates in the vermis, the central nuclei and the cortex of the cerebellar hemisphere. It is derived primarily from the **inferior olivary nucleus** (23–81) on the opposite side, but does have some input from the ipsilateral olive.

(3) The **vestibulocerebellar fasciculus** (15–81) ends in the **fastigial nucleus** (11–81), (4–82) and the cortex of the vermis. It is derived from the ascending branches of the superior, lateral, inferior, and medial vestibular nuclei. It follows a path along the medial aspect of the inferior peduncle and also contains fibers from the spinal nucleus of the trigeminal nerve [V]. (4) The **dorsal external arcuate fibers** (20–81) are derived from the lateral cuneate nucleus of the same side.

The **middle cerebellar peduncle** (7–80), or *brachium pontis,* is made up of three major fasciculi. All are derived from the cells of the nuclei of the pons on the opposite side and terminate in the cortex of the cerebellum. The *inferior fasciculus* constitutes the most caudal transverse fibers of the pons. It is partially covered by the superior fasciculus and terminates in the folia of the undersurface close to the vermis. The *superior fasciculus* consists of the more rostral transverse fibers of the pons. It is the most superficial of the three fasciculi and proceeds in a lateral and backward direction, ending primarily in the inferior surface of the cerebellar cortex. The *deep fasciculus,* as the name implies, includes the deep transverse fibers of the pons. It is initially covered by the superior and inferior fasciculi but does emerge on the medial portion of the superior. Its fibers terminate in the anterior portion of the cerebellar folia. The deep fasciculus covers the fibers of the inferior peduncle.

The **superior cerebellar peduncle** (4–80), or *brachium conjunctivum,* is the smallest and most medial of the three peduncles. It contains most of the efferent fibers of the cerebellum that originate in the gray matter of the cerebellum, specifically, the **globose nuclei** (14–81), (5–82), the **dentate nuclei** (15–79), (12–81), (8–82), and the **emboliform nuclei** (13–81), (6–82). From these nuclei the fibers proceed through the lateral walls of the fourth ventricle and enter the depths of the mesencephalon at the level of the inferior colliculi. The fibers then turn in a rostral direction and terminate in the **red nucleus** (12–79), (8–81) and the adjacent reticular substance of the opposite side. Some of the fibers cross the midline and proceed to the ventral nucleus of the **thalamus** (3–79).

Most of the fibers of the ventral spinocerebellar tract go through the superior cerebellar peduncle to reach the anterior portion of the vermis and the fastigial nuclei.

Figure 80

The cerebellar peduncles.

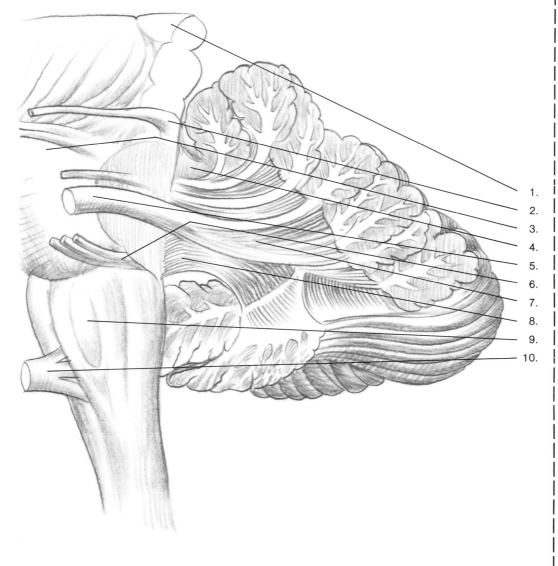

1. *Superior colliculus
2. Trochlear nerve IV
3. Pons
4. *Superior cerebellar peduncle
5. Trigeminal nerve V
6. *Acoustic nerve VIII
7. *Middle cerebellar peduncle
8. *Inferior cerebellar peduncle
9. Inferior olive
10. Hypoglossal nerve XII

Figure 81

Schematic of major connections between the cerebellum and other portions of the brain.

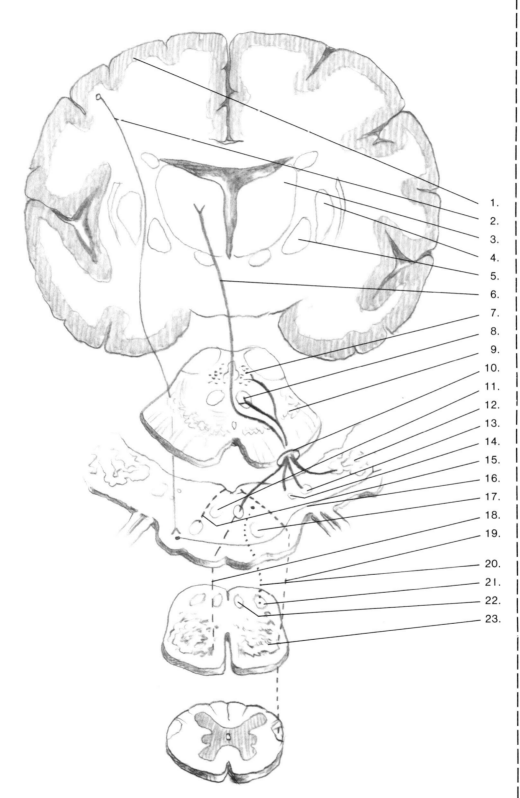

1. Frontal cortex
2. Frontopontile tract
3. *Thalamus
4. Putamen
5. *Globus pallidus
6. Cerebellar thalamic tract
7. Midbrain reticular system
8. Red nucleus
9. Substantia nigra
10. Cerebellorubral tract
11. *Fastigial nucleus
12. Dentate nucleus
13. Emboliform nucleus
14. Globose nucleus
15. Vestibulocerebellar fasciculus
16. Pontocerebellar tract
17. Vestibular nucleus
18. Olivocerebellar fasciculus
19. Dorsal spinocerebellar fasciculus
20. Dorsal external arcuate fibers
21. Nucleus cuneatus
22. Nucleus gracilis
23. Inferior olivary nucleus

THE CEREBELLAR NUCLEI

Four masses of gray matter are located deep in the white portion of the cerebellum on each side. The **fastigial nuclei** (4–82) lie close together on either side of the midline. These are the oldest of the cerebellar nuclei (Fig. 82) and are in a position near the roof of the fourth ventricle. The lateral portion contains relatively large multipolar cells, while the medial section is composed of smaller cells. The fastigial nucleus is sometimes referred to as the *nucleus tecti* because it lies just above the fourth ventricle, separated from it by the fibers of the inferior cerebellar commissure. The shape of this nucleus is spheroidal. It is large enough to be identified in gross sections.

Just lateral to the fatigial nucleus are two or three spherical patches of gray substance that constitute the **globose nucleus** (5–82). The cell structure of the globose nucleus is similar to that of the fastigial nucleus.

Lateral to the globose nucleus near the hilus of the dentate nucleus is the **emboliform nucleus** (6–82). It is wedge shaped but is sometimes difficult to distinguish from the gray matter of the dentate nucleus. In lower forms the emboliform nucleus and the globose nuclei merge to form a single mass of gray matter called *nucleus interpositus.*

The largest and most lateral of the cerebellar nuclei is the **dentate nucleus** (8–82). It is much folded and resembles an empty bag with the mouth, or hilus, opening in a dorsomedial direction. Its appearance is similar to the inferior olivary nucleus of the medulla and was referred to by early anatomists as the cerebellar olive. This nucleus is much larger in humans and the great apes than it is in other mammals and can only be distinguished as a separate nucleus at the mammalian level. The outer surface of the dentate nucleus is formed by a fiber capsule composed of medulated fibers that pass into the nucleus before breaking up into multiple branches.

The internal nuclei of the cerebellum receive fibers from all parts of the cerebellar cortex. Most of them are derived from the long-axoned **Purkinje cells** (1–83). The fastigial nucleus receives fibers from the anterior and posterior portions of the vermis. The area around the vermis projects fibers to the globose and emboliform nuclei on the same side. Some fibers from the same region also project to the cells of the dentate nucleus.

The dentate nucleus, the globose nucleus, and the emboliform nuclei all receive fibers from the lateral hemisphere. All of the internal nuclei also receive fibers from the inferior olive of the medulla.

Figure 82

The internal nuclei of the cerebellum.

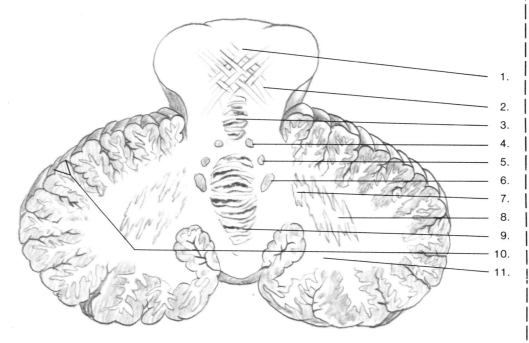

1. Decussation of superior
    cerebellar peduncle
2. *Superior cerebellar peduncle
3. Lingula
4. *Fastigial nucleus
5. Globose nucleus
6. Emboliform nucleus
7. Hilus of dentate nucleus
8. Dentate nucleus
9. Vermis
10. Folia of cerebellum
11. Medullary substance

THE CEREBELLAR CORTEX

The cerebellar cortex is organized in such a way that relatively small stimulation of limited duration is sufficient to produce and perpetuate a large amount of sustained neural cerebellar activity. This phenomenon has been referred to as *avalanche conduction* and can be best understood after examining the cellular structure of the cerebellar cortex (Fig. 83).

The cortex of the cerebellum, when compared with the cerebral cortex, is relatively simple. Unlike the cerebral cortex, it is uniform throughout. It can be readily divided into three well-defined zones within which are five distinct kinds of cells.

The outermost zone is the **molecular layer** (4–83). It contains relatively few cell bodies and is composed primarily of the dendritic processes of the **Purkinje cells** (1–83), the branches of the axons derived from the granule cells, and many of the climbing fibers that twine about the dendrites of the Purkinje cells. The punctate appearance of a transverse section of the cortex results from the fact that the axons run parallel to the folia and are thus severed in a cross section. The molecular layer contains two types of cells, the **outer stellate neurons** (2–83) and the deeper stellate neurons known as **basket cells** (13–83). The outer stellate cells are sparsely scattered through the outer two-thirds of the molecular layer. Their cell bodies are small and star shaped with short, thin dendrites and unmyelinated axons. They synapse with the Purkinje-cell dendrites. The basket cells are located deep in the molecular layer near the cell bodies of the Purkinje cells. The basket cells send out elaborate axonal processes that make contact with the dendrites of the Purkinje cells. A single Purkinje cell may have synaptic connections with many basket cells, and a single basket cell may make contact with many Purkinje cells.

The **Purkinje cells** (1–83) are arranged in a single sheet or layer between the molecular layer above and the granular layer below. The cell bodies are flask shaped with an elaborate arborization emanating from the neck of the flask and penetrating the molecular layer. The dendritic arborization of the Purkinje cells is unusual in that it is flat or fan shaped like an espaliered fruit tree rather than like the branches of a normal tree. The dendrite fan spreads out in a plane at a right angle to the long axis of the folium. Thus, a cross section of the folium will show the full arborization, while a longitudinal section shows only a small portion of the dendritic processes. The axon of the Purkinje cell comes from the cell body just opposite the dendrite, becomes myelinated immediately, and proceeds through the granular layer into the cerebellar white matter. Most of them make contact with the deep cerebellar nuclei. The axons of the Purkinje cells also give off collaterals that go in the opposite direction and end in the molecular layer in conjunction with other Purkinje cells that are located nearby. Thus, the activation of a small group of Purkinje cells results in the recruitment of many more.

The **granular layer** (7–83), which lies between the sheet of Purkinje cells and the white substance, is composed of a remarkable number of **granule cells** (9–83). It has been estimated that there are from 3 to 7 million granule cells per cubic millimeter of granular tissue. Each granule cell sends a single axon into the molecular layer, which divides into two branches in the shape of a "T." These branches run parallel to the long axis of the folium through the dendritic arborizations of the Purkinje cells with which they make synaptic contact. This arrangement also provides for the recruitment of large numbers of cells after the activation of a relatively small number.

It has been estimated that a single Purkinje cell in the cerebellar cortex of the monkey may have as many as 60,000 dendritic spines and that between 200,000 and 300,000 fibers run through that arborization. The granule cells give off from three to five dendritic branches that have clawlike endings and synapse with the mossy fibers.

The upper portion of the granular layer also contains **Golgi type II cells** (5–83). The arborizations of these cells, unlike those of the Purkinje cells, resemble a full tree. They extend into all portions of the cortex but are most predominant in the molecular layer. Also within the granular layer are **glomeruli** (8–83), which are complex synaptic structures composed of mossy fibers, dendrites from granule cells, and dendritic and axon processes of the Golgi cells.

The cerebellar cortex also contains three types of fibers. These include the **axons of the Purkinje cells** (1–83), the **climbing fibers** (12–83), and the **mossy fibers** (11–83). The climbing fibers are relatively fine, penetrating the molecular layer to attach themselves to the dendrites of the Purkinje cells in the manner of a climbing vine. The mossy fibers are coarse and send out a large number of branches that end in the granular area and terminate in assocation with the granule cells.

Figure 83

Schematic of the cellular structure of the cerebellar cortex.

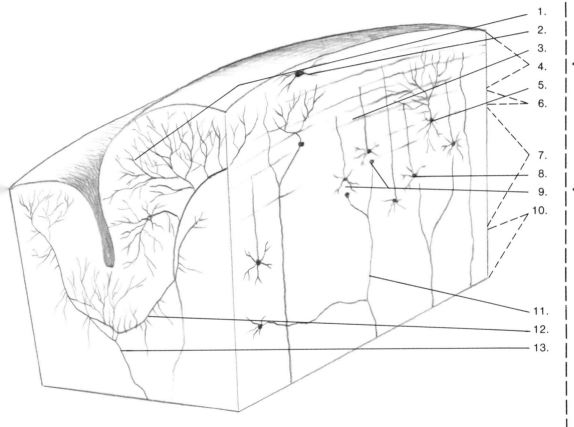

1.	Purkinje cell
2.	Outer stellate neuron
3.	Parallel fibers
4.	*Molecular layer
5.	Golgi type II cell
6.	Purkinje cell layer
7.	Granular layer
8.	Glomerulus
9.	*Granule cell
10.	White matter
11.	Mossy fiber
12.	Climbing fiber
13.	Basket cell

REFLEXES

The extrapyramidal system and the cerebellum are responsible for many of the reflexive postural adjustments that permit voluntary movements to occur smoothly. The behaving individual need not attend to these automatic adjusting mechanisms, which may function better in the absence of conscious awareness. The poet was aware of this phenomenon when he wrote:

> The centipede was happy quite
> Until the frog in fun,
> Said "Pray, which leg comes after which?"
> This raised her mind to such a pitch
> She lay distracted in the ditch
> Considering how to run.

Those reflex mechanisms that involve the cerebellum and the extrapyramidal system are referred to as suprasegmental reflexes. There is also an entire set of reflex mechanisms that can function without input from the brain and are restricted to the spinal cord. Spinal reflexes free of control from higher centers can be studied in animals and humans in which the spinal cord has been severed.

Spinal reflexes occur as relatively invariable patterns of response that are similar for particular kinds of stimuli. Reflexes are innate and consequently do not require any learning or prior experience for their manifestation. They have been selected out during the evolutionary process and are adaptive. The most obvious example of an adaptive reflex is the withdrawal movement that occurs on the application of a painful stimulus to the hand or foot. The intact individual will have an awareness of the pain because the pain messages are also sent to the brain. However, the withdrawal will occur whether the subject is aware of the pain or not, as in the case of a cordotomy (Section 51) in which the pain fibers in the cord have been lesioned.

Postural reflexes are constantly operating in the intact organism. They tend to be long-lasting and sustained reactions. As a result of the postural reflexes, the individual is supported in a standing position with the trunk erect and the head held up.

Phasic reflexes are relatively brief, specific movements that occur when a given stimulus is presented. Phasic reflexes are familiar to all people and include such behaviors as blinking when something comes close to the eye and withdrawal from painful stimulation.

Reflexes may also be classified on the basis of the level of the nervous system involved. Those contained within a single spinal segment are referred to as *segmental reflexes*. Those involving more than one segment are *intersegmental reflexes* and, as previously indicated, reflexes involving the brain are the *suprasegmental reflexes*.

The **stretch reflex,** also called the *myotatic reflex,* involves the contraction of a muscle when it is stretched or lengthened. The familiar patellar reflex is a good example (Fig. 84a). If one sits in a position so that the foot hangs freely and the tendon just below the kneecap is lightly tapped, the **quadriceps muscle** (2–84) will contract with the result that the foot will kick out. The tendon tap results in a lengthening or stretching of the quadriceps muscle with the resultant activation of **muscle spindles** (5–84), which are specialized receptors sensitive to muscle stretch. These receptors lie parallel with the muscle fibers and are interspersed among them. The afferent fibers connected to the muscle spindles enter the **dorsal nerve roots** (4–84) and proceed through the gray matter on the same side of the cord to synapse directly with a **motor neuron** (6–84). This neuron leaves the spinal cord through the ventral root and returns to the quadriceps muscle. Thus, the activation of the muscle spindle results in muscular contraction through a two-neuron arc. Most reflexes involve one or more internuncial neurons between the sensory and motor neurons.

The patellar reflex is an example of a phasic stretch reflex. However, the stretch reflex is also postural and is important as an antigravity mechanism. Under normal conditions, the force of gravity acting on the body produces a tendency for the limbs to buckle. This buckling produces a stretching of the extensor muscles and brings the myotatic reflex into play, with the resultant maintenance of an upright posture.

The simplicity of the reflex mechanism just described is deceiving. The total reflex is much more complex. The afferent neuron also synapses with **ascending fibers** (1, 9–84), which send impulses to the brain that provide the potential for awareness of the state of the muscles. It also synapses with an **inhibitory neuron** (8–84), which innervates the opposing muscle. The function of the inhibitory neuron is to reduce the tone of that muscle with the result that the movement will be free, smooth, and unopposed. The inhibition of antagonistic muscles is common in the motor system and is referred to as *reciprocal innervation*.

(Continued)

Figure 84

Reflexes.

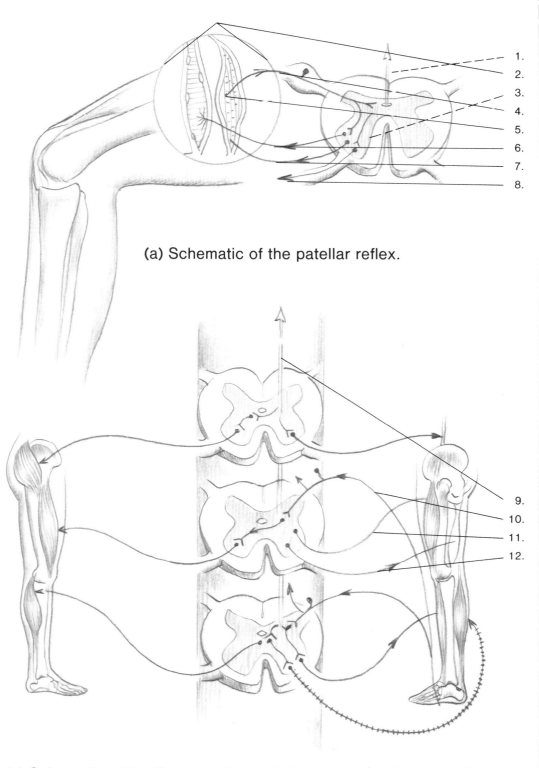

1. Ascending fiber
2. Quadriceps muscle
3. Internuncial neuron
4. Dorsal nerve root
5. Muscle spindle
6. Motor neuron
7. Gamma efferent
8. Inhibitory neuron to flexor

(a) Schematic of the patellar reflex.

9. Ascending fiber
10. Afferent fibers
11. Inhibitory neurons
12. Efferent neurons

(b) Schematic of the flexion reflex and the crossed-extension reflex.

The complexity of the stretch reflex is further increased by the **gamma efferents** (7–84). These motor fibers, which have cell bodies in the anterior horn of the spinal cord, have a relatively small diameter and innervate the specific muscle fibers to which the muscle spindle receptors are attached. Activation of the gamma fibers results in the contraction of those fibers that make the spindle more sensitive to stretching. As a result, the amount of reflex muscle tone fluctuates as a function of changing circumstances. The gamma efferent fibers are activated by descending neurons that probably originate in the anterior portion of the cerebellum.

The *flexion reflex* is a relatively primitive phasic reflex in which the arm or leg is withdrawn from an external stimulus toward the body. The adequate stimulation for the elicitation of this reflex is pain or strong pressure. It is obviously adaptive in that it provides for immediate escape from noxious stimulation. In the lower limbs, the flexion reflex is combined with the *crossed-extension reflex*. When the flexor muscles of one leg are activated and the leg is withdrawn, there is a reflex activation of the extensors in the opposite leg. This prevents the individual from falling when the stimulated leg is lifted from the ground. The flexion and crossed-extension reflexes are diagrammed in Figure 84b.

Although the diagrams illustrating the reflexes are simple, it must be remembered that a very large number of neurons and muscle fibers are involved in even the simplest reflex action. Further, the reflex always occurs against a background of postural adjustments and other activities in the nervous system. As a result, reflexes have a dynamic quality and are not completely predictable on the basis of the stimulus input even under carefully controlled laboratory conditions.

BRODMANN'S AREAS OF THE HUMAN CORTEX

There is considerable variability in the structure of different portions of the cerebral cortex, and they differ in a number of ways. The thickness of the cortical layer is not uniform. The areas differ in the numbers of afferent and efferent fibers. There is variability in the density of the cellular layers and in the distinctness, position, and distribution of the white striae.

Six different cellular layers can generally be identified in the cytoarchitecture of the cortex (see Fig. 86b). Portions of the cortex in which these layers can be readily identified are referred to as *homotypical*. The major portions of the frontal, parietal, and temporal lobes are homotypical. In some cortical areas, however, the six layers are not well delineated. These areas are identified as *heterotypical*. Some authors identify a third classification of cortex called *koniocortex*, in which the cells are small and closer together as found in some portions of the sensory areas.

The differences in cortical structure are sufficiently obvious that it is possible to identify as many as 28 areas with the naked eye. Careful histological studies have resulted in a number of different maps of the cortex that specify the particular areas. The initial study by Campbell in 1905 specified 20 cortical fields. Later investigators have designated 50, 109, and as many as 200 separately identifiable areas. Even some of those 200 areas have been further divided into subareas. To complicate matters further, the various investigators do not always agree. They concur on the more obvious differences, but the finer distinctions are disputed.

Somewhat over fifty years ago Brodmann's extensive studies of the cytoarchitecture of the cortex resulted in the cortical map reproduced in Figure 85. Over the years this chart has become the basic reference for the identification of cortical areas. The numbering system does not have any functional significance. It is based generally on the order in which the different areas were studied. As the diagram shows, there is no relationship between the specified cortical areas and the gyri. A given type of cortex may include a portion of a gyrus or several gyri, and the areas are not delimited by the cortical sulci. In some portions of the cortex the differences between the identified areas are clear, abrupt, and easily specified. In others, the changes are quite gradual, with the result that the boundaries are somewhat arbitrary.

The Brodmann cortical map is reproduced here in order for you to develop a general familiarity with it. It is not suggested, however, that time be devoted to learning each number on the map. In the usual course of study, most of the numbers will never be used. However, as you encounter references to the cortical areas in the literature, refer to the illustration and look them up. Some of the areas will, in time, become quite familiar. It will be recognized, for example, that area 4 is the primary motor area, that area 17 is the primary visual area, and that area 6 can be designated as the premotor area.

Figure 85

The Brodmann cortical map.

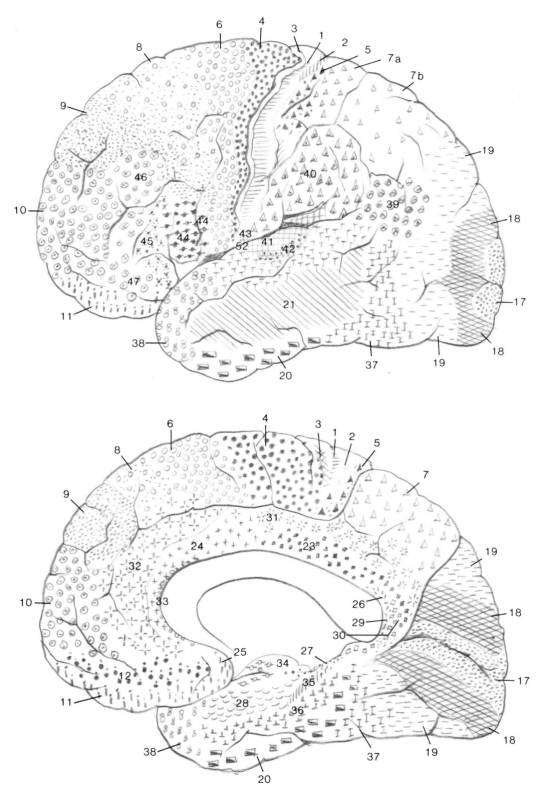

THE STRUCTURE OF THE CEREBRAL CORTEX

The cerebral cortex of the human contains in the neighborhood of 14 billion cells. Although the variability in shape and form is extremely wide, it has generally been agreed that it is reasonable to divide the cortical cells into five types that are readily recognized (Fig. 86a). The most common cell type occurring throughout the cortex is the **pyramidal cell** (1–86). As the name indicates, it is triangular in shape, and the axon extends from the base of the triangle and sends off collaterals that ramify in the adjacent cortex. They range in size from small to giant. Small cells are from 10 to 20 microns, medium from 20 to 25 microns, large from 30 to 35 microns, and the giant size run from 45 to 50 microns. The giant pyramidal cells are called *Betz cells* and are found in the precentral gyrus in the central fissure and on its border. The pyramidal cell has two types of dendrites. The apical dendrite is an extension of the cell body directed toward the surface of the cortex. Smaller dendrites are attached to the sides and the base of the pyramid.

The **granule cells** (2–86) appear to have rounded cell bodies in Nissl preparations. In silver preparations, however, they are star shaped and are thus sometimes referred to as *stellate cells.* Nissl and silver preparations are techniques used in the preparation of materials for microscopic study. These cells are generally quite small with short axons that send off numerous collaterals and terminate on nearby cells. The dendrites of these cells are richly branched and are more complex in higher animal forms. Stellate cells are found throughout the cortex but are particularly abundant in the second and fourth cortical layers.

The **cells of Martinotti** (3–86) are small, multipolar cells present in all areas of the cortex in small numbers. The dendrites from these cells spread out from the cell body in all planes. The axon extends toward the surface of the cortex, sending out collaterals along the way and branching extensively in the outermost layer of the cortex. These cells are associative in function.

The **horizontal cells of Cajal** (4–86) have long horizontally directed dendrites. The myelinated axons may be long or short and run parallel to the surface of the cortex. The cell bodies are small and fusiform or spindle shaped. The horizontal cells of Cajal occur only in the outermost layer of the cortex.

The **polymorphous cells** (5–86), sometimes referred to as *fusiform cells,* are found primarily in the deepest layer of the cortex. Their axons penetrate the subjacent white matter. The long axis of the polymorphic cell body is usually perpendicular to the surface of the cortex.

The nerve fibers of the cortex include axons and dendrites of the cortical cells as well as the afferent fibers that enter the cortex to synapse with the cortical cells. A significant number of the fibers within the cortex run in radial bundles that descend into the white matter from the cortical cells or ascend from the subjacent white matter into the cortex. However, the collaterals and terminal branches of the axons of many of the cortical cells collect into several strata of tangential fibers. These strata form light and dark bands that can be seen in fresh cortical tissue with the naked eye. There is a very thin layer of **tangential fibers** (6–86) in the most superficial layer of the cortex. There are also two clearly defined bands of white tangential fibers called the **outer lines of Baillarger** (10–86) and the **inner lines of Baillarger** (12–86).

The cellular arrangement in the cortex is clearly in the form of a series of laminations. In most parts of the cortex, with the exception of the olfactory areas, six relatively clear and distinct layers can be discerned (Fig. 86b). The layers differ on the basis of the types of cells, the cellular arrangement, and the cell density.

The **molecular layer [I]** (7–86), also called the *plexiform layer,* is the most superficial and contains relatively few nerve cells. The cells present are the horizontal cells of Cajal and some granule cells. The cells from the deeper areas send axons and dendritic ramifications to the molecular layer, forming a complex neuronal plexus that constitutes the superficial band of tangential fibers.

The **external granular layer [II]** (8 86) contains a large number of small pyramidal cells that send fibers to the white matter. Because of the pyramidal cells this strata is sometimes referred to as the *layer of small pyramidal cells.* The external granular layer also contains granule cells and some of the cells of Martinotti.

The **external pyramidal layer [III]** (9–86) can be conveniently divided into two substrata. The deepest area consists primarily of the large pyramidal cells, while the more superficial substrata contains pyramidal cells of medium size. Granule cells and cells of Martinotti may also be found in this layer.

The **inner granular layer [IV]** (11–86) consists essentially of closely packed granule cells with small pyramidal cells scattered among them. The **internal pyramidal layer [V]** (13–86), or *ganglionic layer,* contains the largest

(Continued)

Figure 86

The structure of the cerebral cortex.

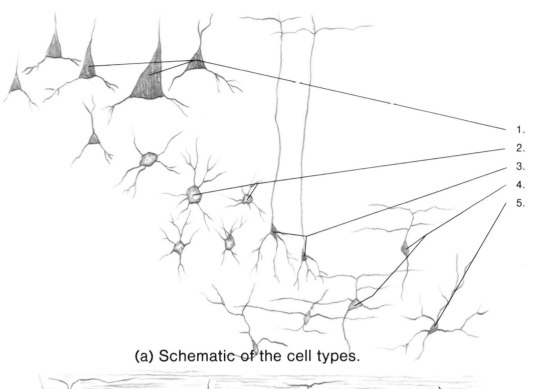

(a) Schematic of the cell types.

1.	*Pyramidal cells
2.	*Granule cells
3.	Martinotti cells
4.	Horizontal cells of Cajal
5.	*Polymorphous cells

(b) Schematic of the six cellular layers.

6.	Tangential fibers
7.	*Molecular layer I
8.	*External granular layer II
9.	External pyramidal layer III
10.	Outer lines of Baillarger
11.	Inner granular layer IV
12.	Inner lines of Baillarger
13.	*Internal pyramidal layer V
14.	*Polymorphic cell layer VI

of the pyramidal cells. In the motor area, this strata contains the giant *cells of Betz.* The axons of these cells in the motor area constitute the corticobulbar and corticospinal tracts. The apical dendrites from these cells proceed toward the superficial stratum of the cortex and ramify in the molecular layer. Medium-sized pyramidal cells as well as granule cells and cells of Martinotti are also found in layer V.

The **polymorphic cell layer [VI]** (14–86), or *layer of fusiform cells,* is the sixth and innermost stratum of the cortex. It is composed of numerous spindle-shaped and angular cells that send axons into the white matter immediately below. The dendrites of these cells project into the fourth and fifth layers.

SECTION 65

THE CORTICAL FUNCTION

The areas indicated in Figure 87 as serving particular functions are, at the very best, approximations. Although there is reasonably precise localization of function in the primary sensory and motor areas, it is possible to elicit minor motor movements outside designated motor areas, and sensations can be produced by electrical stimulation outside the primary sensory areas. For example, Penfield found in his study of the unanesthetized brain of neurosurgical patients that 18 percent of the somesthetic responses for the facial area were elicited by stimulation of the *precentral* region adjacent to the facial area in the postcentral gyrus. Stimulation of the *precentral* gyrus also elicited 27 percent of the arm and leg sensations. The areas not associated with primary motor or sensory experience can be even less reliably specified.

In this section four types of function will be considered: sensory, motor, suppressor, and association. In order to avoid overloading the diagram, those portions of the cortex concerned with language functions will be considered in Section 66.

Although the relationship between the Brodmann cortical areas and specific functions is considerably lacking in precision, the Brodmann cortical map provides a very useful method of locating specific areas under discussion. You should refer frequently to the diagram of Brodmann areas Fig. 85 when studying this section.

The specific primary projection areas have been covered in earlier sections on each of the sensations and will only be briefly considered here. The **primary visual area** (11–87) includes area 17 in the area of the calcarine fissure. Destruction of that area in humans results in total cortical blindness. The primary reception area for **audition** (14–87) includes areas 41 and 42, and probably area 52. Areas 41 and 42 cannot be seen completely unless the temporal lobe is retracted from the parietal lobe. Areas 3, 1, and 2 serve as the **somesthetic projection areas** (3–87). Lesions in this portion of the cortex result in a loss of the finer aspects of tactile recognition. The cortical location of the **gustatory area** (16–87) is in some debate, but the best evidence seems to place it at the base of the postcentral gyrus. Finally, **olfaction** (31–87) appears to be appreciated in the area of the uncus.

The **primary motor area** (2–87) is located just anterior to the central fissure and consists of Brodmann's area 4. Stimulation of this portion of the cortex results in a pat-

tern of movement depending on the particular part stimulated. The specific movements can be predicted on the basis of the motor homunculus in Figure 78.

Rostral to area 4 is area 6, which is the **premotor area** (6–87). Stimulation of area 6 results in a pattern of movement similar to that obtained with stimulation of a comparable portion of area 4, but the intensity of stimulation must be much greater.

Frontal eye fields (15–87), which contribute to the control of eye movements, are located in the inferior portion of area 8. A second area for the control of eye movements is the **occipital eye field** (12–87), which is located in a portion of area 19.

It has been repeatedly shown in animal experimentation that electrical stimulation of particular areas of the cortex results in the blocking of some types of ongoing activity. Those portions of the cortex are known as **suppressor areas** (1, 4, 8, 13, 19, 21, 24, 29–87). The **precentral suppressor area** (1–87) has been verified in humans, but some doubt remains about the specific location of other suppressor areas in the human brain. Figure 87 shows those portions of the cortex where suppressor functions might be expected on the basis of experimentation in subhuman primates.

Those portions of the cortex in the immediate vicinity of the primary sensory areas are involved in the integration of specific sensations into meaningful patterns and are referred to as association areas. It is in these areas that primary sensations are built into perceptions so that the individual is able to recognize and attach meaning to complex sensory input. Disturbances in the association areas result in a loss of the ability to perceive meaningful relationships. Such losses are called *agnosias*. Depending on the portion of the cortex involved, agnosias can include tactile, visual, or auditory stimuli. Agnosias that concern the highly complex associations involved in the use of language are called *aphasias* and are discussed in the next section.

The **tactile association area** (5–87) lies just behind the somesthetic cortex in Brodmann's areas 5 and 7. Lesions in this part of the cortex result in *astereognosis*. With this condition patients are unable to recognize familiar objects by touch alone. They may, for example, be able to tell you that a particular object is hard and cold, but will not be able to identify it as a pair of scissors until they look at it.

The **visual association area** (10–87) is found in the

(Continued)

Figure 87

Schematic of the functional areas of the cortex.

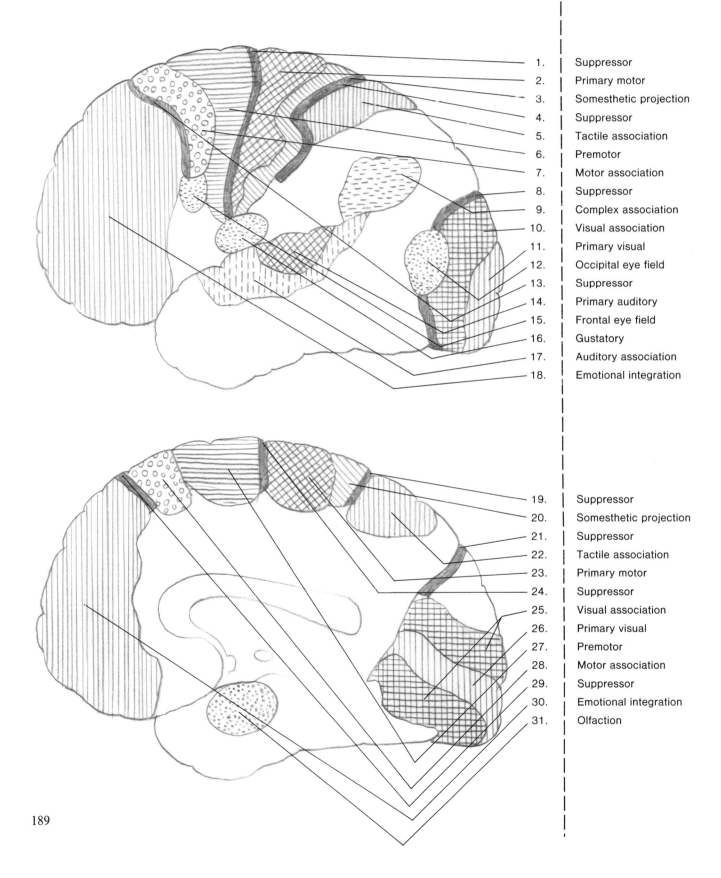

1. Suppressor
2. Primary motor
3. Somesthetic projection
4. Suppressor
5. Tactile association
6. Premotor
7. Motor association
8. Suppressor
9. Complex association
10. Visual association
11. Primary visual
12. Occipital eye field
13. Suppressor
14. Primary auditory
15. Frontal eye field
16. Gustatory
17. Auditory association
18. Emotional integration

19. Suppressor
20. Somesthetic projection
21. Suppressor
22. Tactile association
23. Primary motor
24. Suppressor
25. Visual association
26. Primary visual
27. Premotor
28. Motor association
29. Suppressor
30. Emotional integration
31. Olfaction

occipital lobe primarily in Brodmann's area 18. Damage to area 18 results in visual agnosia, in which individuals are unable to appreciate the meaning of visual stimuli. They may be able to see a chair, as evidenced by their tendency to avoid walking into it, but they cannot name it or explain its use.

The **auditory association area** (17–87) is located in the temporal lobe including Brodmann's area 22. Lesions result in word deafness and other auditory agnosias, which may include the inability to interpret such familiar sounds as the ring of the telephone or favorite musical compositions.

A **complex association area** (9–87) that involves higher-order nonverbal integration is located in the parietal lobe including parts of Brodmann's areas 39 and 40. Damage to this part of the cortex results in the bizarre disorder known as *body-scheme failure*. The individuals are capable of recognizing a hand and identifying it as such, but they are unable to appreciate the fact that their own hand belongs to themselves.

There is a **motor association area** (7–87) in which voluntary motor movements are integrated. Damage to this area results in a loss of the ability to put basic voluntary movements together to produce skilled behaviors. This disorder is called *apraxia*. One result of this damage may be *agraphia,* the inability to write. Apraxia may also result from lesions in the somesthetic association area of the parietal lobe. In this case the subject is unable to carry out a sequence of movements even though each single movement may be performed adequately.

The **prefontal areas** of the cortex, those rostral to areas 6 and 8, have been extensively studied since the advent of the surgical procedure of prefrontal lobotomy in 1935. Its function is difficult to characterize. Much of the prefrontal area receives fibers from the dorsomedial nucleus of the thalamus. Thus, impulses are relayed to the cortex from the autonomic areas within the hypothalamus. The prefrontal areas are concerned with the integration of emotional responses and, in particular, learned emotional reactions. Damage to the prefrontal area does not result in a deficit in intelligence as measured by the common intelligence tests. However, it does result in a loss of what has been called *biological intelligence,* derived from a series of tests by Halstead that measured such variables as critical flicker-fusion frequency, memory of time-sense, tactual form-board performance, and others.

SECTION 66

APHASIAS

Language disorders that result from brain damage or disease are called aphasias. Figure 88 is a schematic representation of the types of aphasias and their relation to the particular areas of the cortex frequently involved. Diagrams of this type are necessarily simplifications and should be interpreted with caution. Language is a complex symbolic activity, and no single aspect of it is represented in a small cortical area. Further, all of the cortical areas involved in language are richly interconnected. A lesion in one specific area will therefore result in a dysfunction in the ability to communicate that will involve more than one aspect of the language facility.

In general, the aphasias can be divided into motor aphasias of the expressive type and sensory aphasias of the receptive type. In motor aphasia individuals are unable to express themselves verbally. They may know what they want to say and be able to produce a full range of sounds, but they are unable to put those sounds together so that they have meaning. This loss of articulate speech occurs with damage to **Broca's area** (4–88), Brodmann's area 44, in the dominant hemisphere. It is possible to sustain damage to area 44 and still retain the ability to sing or produce rhythmic speech. It is believed that these skills are represented bilaterally in area 45, which is anterior to Broca's area.

Another motor aphasia involves the loss of the ability to write, known as **agraphia** (1–88). This occurs with damage to the cortical area above 44 in a part of Brodmann's area 9. The patients are unable to write even though their general motor coordination is unimpaired.

Destruction of the portions of the temporal cortex in Brodmann's area 22 results in a sensory aphasia in which the patients are unable to understand the spoken word even though general hearing is unaffected. This is called **word deafness** (6–88). In this condition, speech often becomes confused because the individuals are unable to understand what they have just said.

Word blindness, or **alexia** (3–88), is a form of sensory aphasia in which the patients are unable to recognize the printed or written word and are consequently unable to read. They may also manifest *agraphia* because they are unable to read what they have just written. This condition results from destruction of the angular gyrus, Brodmann's area 39.

A particularly severe form of sensory aphasia occurs when both areas 22 and 39 are involved (5–88). The patient is unable to understand either auditory or visual language input. This condition is referred to as *Wernicke's aphasia*.

Figure 88

Cortical areas associated with language functions.

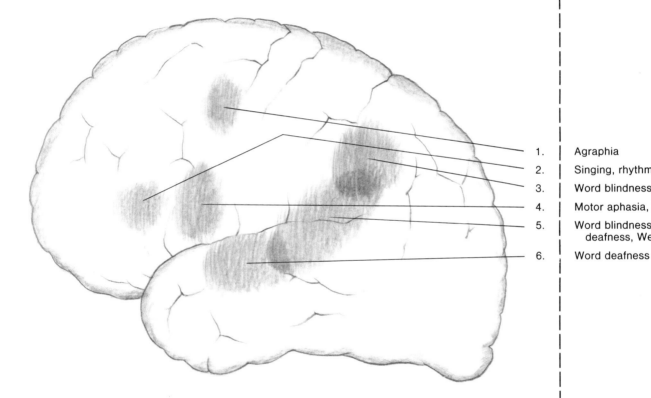

1. Agraphia
2. Singing, rhythmic speech
3. Word blindness
4. Motor aphasia, Broca's area
5. Word blindness and word deafness, Wernicke's area
6. Word deafness

THE RETICULAR ACTIVATING SYSTEM

The reticular system is involved in the entire central nervous system. It is derived from a core of tissue that runs more or less centrally through the entire brain stem from the cell bodies of the central spinal cord to the central portions of the thalamus. The term *reticular formation* is derived from the microscopic appearance of the cellular tissue that makes up the system. With proper staining the cells can be seen to be embedded in a mesh or network of fibers that seem to be randomly distributed, much like veins in a leaf.

Anatomically and functionally the reticular activating system (RAS), diagrammed in Figure 89, can be divided into two major systems, the **ascending system** (11–89) and the **descending system** (13–89). The descending system originates in the pons and the medulla and descends generally through the **reticulospinal tracts** (21–89). Some of the fibers, however, are not clearly separated in the cord. The pontine fibers have their origin in the **oral pontine reticular nucleus** (12–89). These fibers descend in the anterior portion of the cord and except for a few fibers do not cross to the opposite side.

The medullary fibers originate in the **nucleus reticularis gigantocellularis** (16–89) and the **paramedian reticular nuclei** (18–89), which are located dorsal to the **inferior olivary nucleus** (19–89). These fibers descend in the anterolateral portion of the cord. Experimental evidence indicates that the reticulospinal tracts descend the entire length of the spinal cord.

The nuclei just indicated also contribute fibers to the ascending reticular system. In fact, many of the cells of the reticular system at various levels send out fibers that bifurcate, with one collateral ascending and the other descending. At the level of the medulla, the **lateral reticular nucleus** (17–89) contributes fibers to the ascending system that join fibers from the pontine reticular nuclei. These fibers pass through the **central tegmental tract** (9–89) to terminate in the hypothalamus and the **intralaminar nuclei of the thalamus** (3–89).

The midbrain reticular formation is less extensive than that found in the pons, but it includes a significant portion of the medial and lateral gray matter. The **pedunculopontine tegmental nucleus** (6–89) is one of the larger identifiable nuclei associated with the midbrain reticular system. It lies just lateral to the central gray area and ventral to the nucleus of the **inferior colliculus** (5–89). These reticular nuclei in the mesencephalon also project fibers to the intralaminar nuclei of the thalamus.

The exact connections of the reticular system from the thalamus to the telencephalon are in some doubt because anatomical and physiological techniques of study sometime yield different results. However, it is clear that there are connections, many of which are reciprocal, between the intralaminar nuclei and most of the cortex. The more-caudal portions of the system also send fibers to the **hypothalamus** (15–29) and thence through the **medial forebrain bundle** (2–89) to the **septal region** (1–89), which is generally considered to be the most rostral extent of the reticular system. Reciprocal connections also exist between the septum and the amygdala and between the septum and the **hippocampus** (8–54).

The RAS receives secondary, or collateral, fibers from all of the sensory systems. Connections from the visual system reach the reticular system through relays in the superior colliculi. Secondary fibers enter the RAS from the auditory and vestibular systems as well as from the major ascending fasciculi in the spinal cord, with the exception of the medial lemniscus, which apparently does not make such connections.

The functioning of the reticular activating system has profound effects on the normal moment-to-moment behavior of the individual, and is critical to consciousness itself. Stimulation of the descending system in the ventromedial portion of the reticular cells of the medulla results in an inhibition of almost all forms of motor activity, including reflexes, muscle-tone reduction, and motor activity elicited by cortical stimulation.

Stimulation of the more-dorsal portions of the reticular substance all the way from the medulla up through the hypothalamic portions and the intralaminar nuclei of the thalamus results in the facilitation of motor movement rather than its inhibition. Again, all types of motor behavior are facilitated. Muscle tone is evidently influenced by the action of the RAS on the **gamma motor neurons** (7–84).

The respiratory response of maximal inspiration can be obtained by stimulation of portions of the nucleus reticularis gigantocellularis in the medulla. Stimulation in a different portion of that nucleus results in a drop in blood prssure. Responses of maximum expiration and significant increases in blood pressure occur when other portions of the medullary reticular formation are stimulated.

The vast array of sensory input to the RAS permits it to function as an arousal system. Any sensory input to

(Continued)

Figure 89

The reticular activating system.

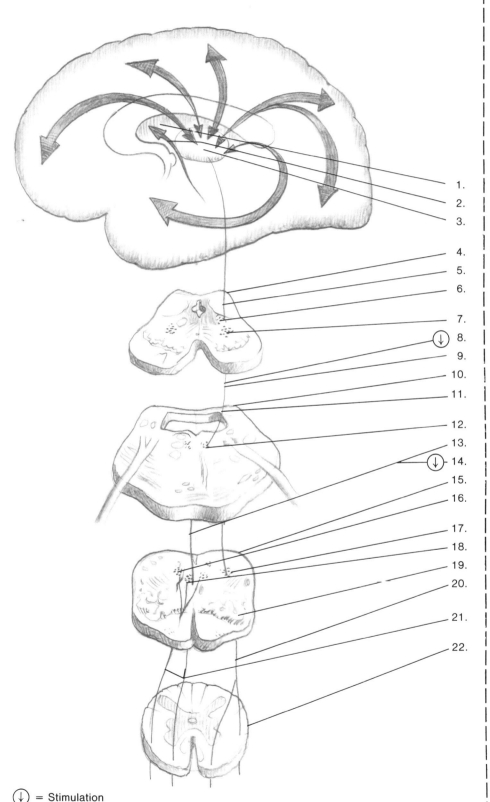

1. Septal region
2. Medial forebrain bundle
3. Intralaminar nucleus of thalamus
4. *Midbrain
5. *Inferior colliculus
6. Pedunculopontine tegmental nucleus
7. Red nucleus
8. Arousal
9. Central tegmental tract
10. Pons
11. Ascending reticular activating system
12. Oral pontine reticular nucleus
13. Descending reticular system
14. Motor activity inhibition
15. Medulla
16. Nucleus reticularis gigantocellularis
17. Lateral reticular nucleus
18. Paramedian reticular nucleus
19. Inferior olivary nucleus
20. Ascending spinoreticulo fasciculi
21. Descending reticulospinal fasciculi
22. *Spinal cord

↓ = Stimulation

the system produces an increase in electrical activity that spreads to other portions of the system. The result of this amplification is that activity in all parts of the nervous system is increased. The cortex is stimulated and the organism is activated or aroused. Thus, a sleeping animal can be awakened by a relatively minor sensory input. The RAS is critical to wakefulness and consciousness. This aspect of the RAS will be covered in more detail in the section on sleep. Since the RAS receives considerable input from the cortex, a general state of arousal can be generated by activity in the cortex itself. Nonconvulsive stimulation of a large number of cortical areas results in activity in the RAS and a generalized activation. The largest number of the fibers from the cortex travel to the nucleus reticularis gigantocellularis in the medulla and to the oral pontine reticular nucleus in the pons.

RESPIRATION

It is obvious that individuals can voluntarily control their breathing. However, they usually do not. Respiration is most generally controlled by a relatively complex reflex mechanism. Further, voluntary control is limited. Voluntary inhibition of breathing until death occurs by asphyxiation is not possible. As soon as consciousness is lost, reflex mechanisms take over and individuals once again begin to breathe involuntarily.

The primary areas involved in the control of respiration are bilateral and lie in the medullary portion of the reticular activating system (Fig. 90). One portion is related to inspiration, while a second is concerned with expiration. It is a misnomer to refer to these areas as centers. They are, in fact, only one part of a complex respiratory system. The areas serving the respiratory function are rather ill-defined and cannot be restricted to particular nuclei. Further, the areas include a variety of unrelated structures that serve other functions. These areas extend from just below the pons as far caudally as the level of the **calamus scriptorius** (50–8). The portion involved in **inspiration** (9–90) is located a few millimeters on either side of the midline just anterior to the **inferior olive** (10–90), while those subserving the function of **expiration** (7–90) are more dorsal and rostral.

When the carbon-dioxide level in the blood reaches a certain concentration, electrical activity in the inspiration area increases. However, it does not appear to be influenced by the blood levels of oxygen. Cells from this area send fibers to the third, fourth, and fifth cervical segments of the spinal cord, where they synapse with those cells that send axons through the **phrenic nerve** (21–90). Fibers also descend to the thoracic segments, where they activate neurons that control the intercostal and abdominal muscles. Thus, sufficient activity in the areas of inspiration produces contractions of the muscles that result in the filling of the lungs.

Within the lungs, at the bifurcation of the bronchioles, are **stretch receptors** (23–90) that are activated by the lung expansion involved in inspiration. Impulses from those receptors travel through the **vagus nerve [X]** (15–90) to enter the **tractus solitarius** (8–90) of the medulla. They are then relayed to the area of expiration, which when activated results in the inhibition of the muscles of inspiration and the activation of those related to expiration. This is known as the *Hering-Breuer reflex.*

If the vagi are cut, eliminating impulses from the stretch receptors, respiration slows and deepens but it does not stop. This is evidently the result of the activity of the so-called pneumotaxic "center" in the pons, which sends fibers to and receives them from the medullary respiratory areas. The pneumotaxic "center" is located at the junction of the pons and the midbrain, probably in the area of the **nucleus locus ceruleus** (4–90), also called the *pigmented nucleus.* Activity in the pneumotaxic "center" results in expiration. If both the pons and the vagi are damaged, inspiration cannot be inhibited and the subject dies of prolonged inspiration.

Reflexive control of the process of respiration can be modified but not completely controlled by voluntary or emotional activation from higher centers. Experimental work has shown that stimulation of the orbital cortex, the anterior hippocampal gyrus including the uncus, and the temporal pole results in the inhibition of respiration. Accelerated respiration typical of that found in various emotional states results from the stimulation of much of the lateral hypothalamus.

The mechanisms already described are involved in normal breathing. However, there is a backup system that probably comes into action only in unusual circumstances, such as during deep anesthesia and other times of oxygen deprivation. Chemoreceptors are located in the **carotid body** (12–90), which is situated at the bifurcation of the common carotid artery, and in the **aortic body** (18–90), which is in the arch of the aorta. These receptors are sensitive to both the oxygen and carbon-dioxide levels in the blood and are less sensitive to asphyxia than the respiratory areas in the medulla. Thus, in times of oxygen deprivation the medullary centers can be driven by impulses from those two bodies. Impulses travel from the carotid body through the glossopharyngeal nerve [IX] to the tractus solitarius and thence to the medulla. Impulses from the aortic body travel through the vagus nerve [X].

Finally, inspiration can be reflexly induced by stimulation of the skin by a sudden slap or application of cold water to the face or body. Impulses from the sensory fibers of the skin ascend through the **lateral spinothalamic tract** (19–90) or the secondary trigeminal fibers or both and are thence transmitted to an inspiration area of the medulla. Thus, many newborn babies are slapped on the bottom to initiate reflex breathing.

Figure 90

Neural mechanisms involved in respiration.

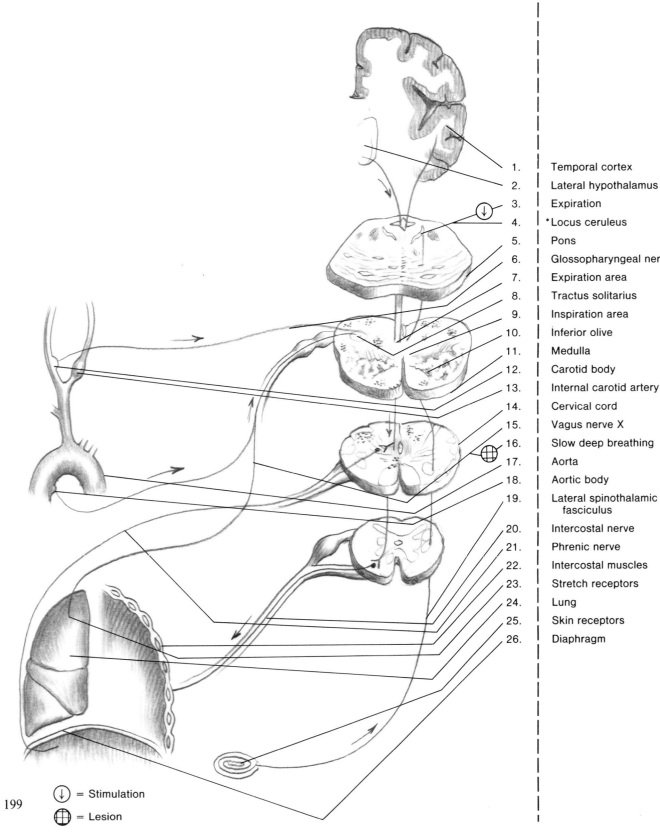

1. Temporal cortex
2. Lateral hypothalamus
3. Expiration
4. *Locus ceruleus
5. Pons
6. Glossopharyngeal nerve IX
7. Expiration area
8. Tractus solitarius
9. Inspiration area
10. Inferior olive
11. Medulla
12. Carotid body
13. Internal carotid artery
14. Cervical cord
15. Vagus nerve X
16. Slow deep breathing
17. Aorta
18. Aortic body
19. Lateral spinothalamic fasciculus
20. Intercostal nerve
21. Phrenic nerve
22. Intercostal muscles
23. Stretch receptors
24. Lung
25. Skin receptors
26. Diaphragm

↓ = Stimulation

⊕ = Lesion

BLOOD PRESSURE

Blood pressure is kept within relatively narrow limits by a complex mechanism involving all levels of the nervous system as well as certain hormonal factors. The basic reflex control is maintained by a pressure-sensitive feedback loop that operates through the medulla oblongata and the autonomic nervous system (Fig. 91). This servomechanism is influenced by hormonal factors, including norepinephrine and epinephrine from the adrenal glands and renin from the kidneys. Psychic and emotional factors influence the higher centers of the nervous system, which may further modify the operation of the basic control system.

Collections of cell bodies that react to changes in pressure are called *baroreceptors*. They are located in the **carotid body** (19–91) at the bifurcation of the **internal carotid artery** (20–91), and in the **aortic body** (23–91), which is in the arch of the **aorta.** The baroreceptors function in a manner similar to a house thermostat and are sometimes referred to as *barostats*. The barostats are set for a given blood-pressure level. When the pressure exceeds that level, the cells are activated and impulses are sent to the pressure-controlling mechanism in the medulla. Impulses from the carotid body travel through the **glossopharyngeal nerve [IX]** (17–91), while those from the aortic body go through the **vagus nerve [X]** (18–91). That particular branch of the vagus is called the **depressor nerve.**

The **depressor area** (15–91) is located in the same general portion of the medulla as the cells that control inspiration (9–90). The depressor cells are not restricted to particular nuclei and are intermingled with cells serving different functions. When depressor cells are activated, they send impulses through the vagus to the heart that have an inhibitory function and produce a slowing of the heart rate. Impulses also flow down the spinal cord and have an inhibitory effect on the sympathetic fibers that control blood-vessel constriction. The consequent dilation of the arterioles and small vessels permits the blood to flow more easily, and blood pressure is reduced.

As the pressure decreases, the baroreceptors become less active and there is a reduction in vagal output. The blood pressure then begins to rise again. As a result of this highly sensitive mechanism, the blood pressure is in a constant state of relatively small fluctuations around a given level.

The medulla also contains a relatively ill-defined **pressor area** (12–91), which when activated results in an increase in blood pressure. It is also in the reticular formation of the brain stem and overlaps somewhat with the **expiration area** (7–90). As with the areas for respiration, it is inappropriate to refer to this portion of the blood-pressure control mechanism as the "blood pressure center." It is only one portion of the complex system for pressure control.

When the pressor area is activated, impulses are sent down the spinal cord to activate the sympathetic nervous system. The general result is the constriction of particular sets of blood vessels, the acceleration of the heart rate, and the stimulation of the adrenal medulla. The latter action results in the output of epinephrine and norepinephrine into the blood stream. These hormones function to enhance and prolong the effects of the sympathetic nervous system. The increased heart rate and vascular constriction result in an increase in blood pressure.

The vasomotor mechanisms in the medulla are subject to influences from the higher centers in the brain that serve psychic and emotional functions. Experimental evidence has shown that there are a variety of areas in the brain that when stimulated electrically have a pressor or a depressor effect.

As a general rule, it is useful to think of the hypothalamus as divisible into an anterior and posterior portion in regard to its influence on autonomic function. Stimulation of the posterior hypothalamus results in activation of the sympathetic system and has been called the *ergotropic zone*. Impulses travel from there to the pressor area of the medulla, and a rise in blood pressure follows. The anterior part of the hypothalamus has been called the *trophotropic zone* and is associated with parasympathetic function. Impulses from this part of the hypothalamus descend to the depressor portion of the medulla with the result that blood pressure is reduced.

Like most general rules of the nervous system, this is somewhat of an oversimplification. Some investigators have found pressor points in the more anterior portions of the hypothalamus, particularly in the lateral portion in the area of the **medial forebrain bundle** (1–34d), and there is an intermingling of pressor and depressor points in other parts of the hypothalamus.

Vasomotor changes also result from stimulation in other brain areas. Blood-pressure reduction has been produced by stimulation of the **preoptic region** (5–34d), the **septum pellucidum** (3–10), and the **dorsal thalamus**

(Continued)

Figure 91

Neural mechanisms involved in blood-pressure control.

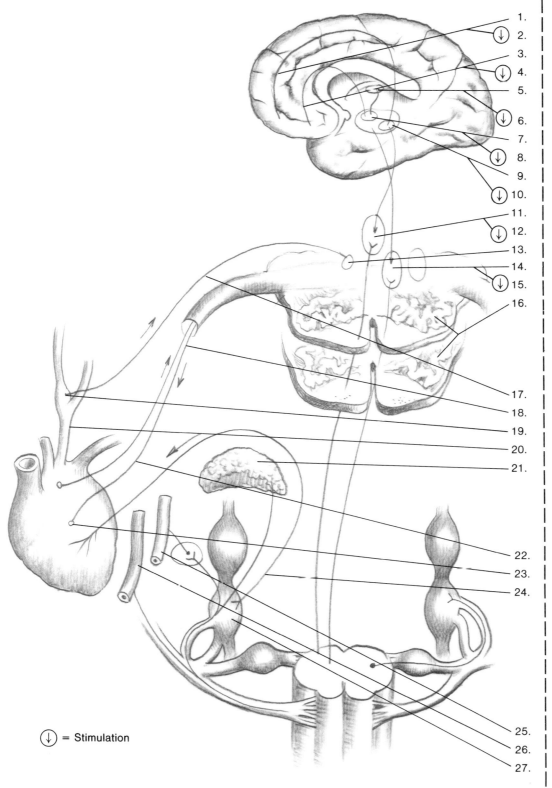

1. Frontal lobe
2. Pressor effect
3. Orbital frontal area
4. Depressor effect
5. Dorsomedial nucleus of thalamus
6. Depressor effect
7. Anterior hypothalamus
8. Depressor effect
9. Posterior hypothalamus
10. Pressor effect
11. Medulla
12. Pressor effect
13. Nucleus of tractus solitarius
14. Medulla
15. Depressor effect
16. Inferior olive
17. Glossopharyngeal nerve IX
18. Vagus nerve X
19. Carotid body
20. Internal carotid artery
21. Adrenal gland
22. Accelerator to heart
23. Aortic body
24. Accelerator to heart
25. Visceral blood vessels
26. Peripheral blood vessels
27. Sympathetic chain

⊥ = Stimulation

(5–91). Stimulation of particular areas of the frontal lobe have resulted in blood-pressure increase and heart-rate acceleration, and the activation of the orbital surface of the frontal cortex may cause an immediate drop in blood pressure.

It is presumed, of course, that these various areas of the brain are involved in the normal day-to-day regulation of blood pressure. Input from the higher centers, particularly those involved in emotional responses, may override the reflexive mechanisms for blood-pressure control and disrupt them to such an extent that a permanent state of hypertension (high blood pressure) ensues. Chronic states of anger, worry, anxiety, and tension result in a continuous activation of the pressor mechanism in the medulla. Under the influence of continued abnormally high pressure, the barostats in the carotid and aortic bodies appear to become reset so that they are reflexly activated to reduce the blood pressure only by levels that are significantly higher than normal for the individual.

TEMPERATURE REGULATION

The temperature of cold-blooded (*poikilothermic*) animals rapidly changes to be congruent with the temperature of the environment. On the other hand, warm-blooded (*homiothermic*) animals, including humans, maintain a reasonably constant internal temperature. Homiothermy is achieved at no small cost and involves widespread physical, chemical, and metabolic processes that include all levels of the nervous system (Fig. 92).

Temperature maintenance is accomplished through two different processes. In the first instance the organism is protected against excessive loss of heat (*hypothermia*), and in the second, heat is dissipated to keep the individual's temperature from going too high (*hyperthermia*).

Hypothermia can be prevented either by decreasing the heat loss, or by increasing the heat production. Heat loss is limited by the constriction of the peripheral blood vessels, which limits the exposure of the blood to the lower environmental temperature. In animals, heat is also conserved by raising the hairs (*piloerection*) to trap more air, which functions as insulation. Although no longer useful in temperature regulation, humans still maintain the mechanism for piloerection. Thus, a reduction in environmental temperature produces gooseflesh. Heat production can be increased by raising the basal metabolic rate through the activation of the thyroid, and by the process of shivering. Shivering involves repeated contractions of the striate muscles throughout the body. These contractions increase the process of oxidation with a consequently greater production and conservation of heat.

The mechanisms involved in the prevention of hyperthermia include dilation of the peripheral blood vessels, sweating, and increased respiration in humans and panting in animals. Peripheral vasodilation provides for the elimination of heat through the process of convection and radiation from the surface of the expanded blood vessels. The evaporation of perspiration reduces the temperature, and rapid respiration produces heat loss through the increased air and water vapor expired through the lungs.

A great deal of clinical data and experimental work have demonstrated that the **hypothalamus** (5–92) is of the utmost importance in the regulation of body temperature. If most of the hypothalamus is rendered nonfunctional through disease, tumor, or experimental lesion, the organism becomes poikilothermic and the body temperature fluctuates with the environmental temperature. When some cell clusters in the hypothalamus are activated, they produce changes that function to reduce temperature, whereas the action of other cell clusters is to increase the body temperature. These two different functions are not restricted to particular hypothalamic nuclei but are distributed in different parts of the hypothalamus. However, the distribution is not random. There is a concentration of the cells concerned with heat dissipation in the **preoptic region** (7–92) and in the anterior hypothalamus. Thus, damage to the heat-dissipation regions results in *hyperpyrexia* (extremely high fever). In humans these "neurogenic fevers" may be caused by localized tumors or by damage to the area during surgery. Stimulation of the preoptic or anterior hypothalamic nuclei results in sweating, peripheral vasodilation, and an increase in respiratory rate. In animals such stimulation produces panting.

The patient with damage to the anterior region of the hypothalamus is no more subject to chilling than the normal individual. The cellular groups concerned with the prevention of heat loss tend to be concentrated in the **posteromedial regions of the hypothalamus** (9–92). Damage to this area results in an inability to mobilize the physiological process for the maintenance of body temperature, with the result that the patient is subject to hypothermia and his temperature is perpetually below normal. Stimulation of this hypothalamic area produces shivering, peripheral vasoconstriction, piloerection, and activation of the adrenal medulla with the release of epinephrine and norepinephrine. There is also an increase in metabolic rate, which is probably due to the increase in thyroid activity. That increase occurs because the pituitary has been induced to put out the thyrotropic hormone that facilitates thyroid function.

Some temperature control is achieved through vasoconstriction produced by local reflexes. The stimulation of **cold receptors** (23–92) in the skin sends impulses to the spinal cord that reflexly activate the **sympathetic system** (24–92), causing blood vessel constriction in the immediate area. Thermal messages also reach the hypothalamic control mechanisms through the **lateral spinothalamic tract** (21–92) and influence the activity in those control-mechanisms. However, the hypothalamus itself contains thermal receptors sensitive to changes in the temperature of the circulating blood, which, of course, reflects body temperature. Experimental work

(Continued)

Figure 92

Neural mechanisms involved in temperature regulation.

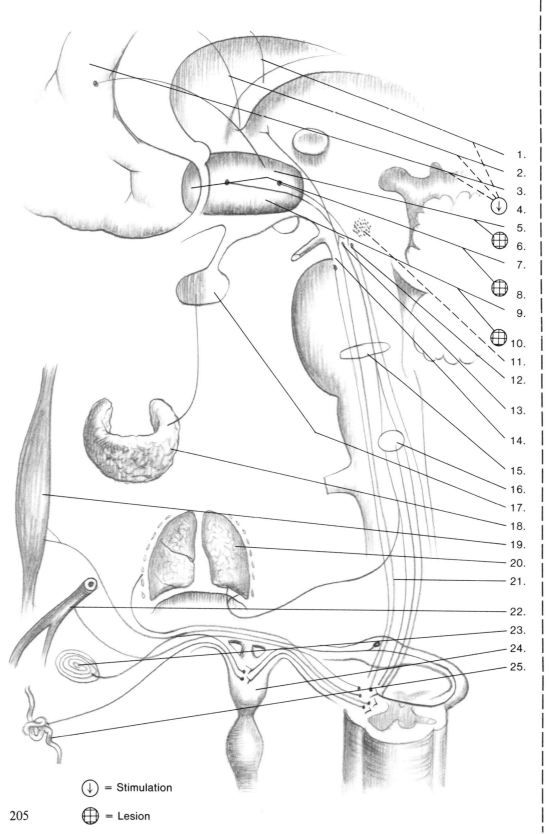

1. From area 4
2. From area 6
3. Cingulate gyrus
4. Vasoconstriction, shivering
5. Hypothalamus
6. Poikilothermia
7. Preoptic region, heat dissipation
8. Hyperthermia
9. Posteromedial hypothalamus, heat conservation
10. Hypothermia
11. Red nucleus
12. Posterior hypothalamic tegmental tract
13. Anterior hypothalamic tegmental tract
14. Ventral mesencephalic tegmentum
15. Tegmentobulbospinal fibers
16. Medulla reticular formation
17. *Pituitary
18. Thyroid
19. Striped muscle, shivering
20. Lungs
21. Lateral spinothalamic fasciculus
22. Peripheral blood vessels
23. Cold receptor
24. Sympathetic ganglia
25. Sweat glands

↓ = Stimulation

⊕ = Lesion

has shown that direct heating or cooling of the hypothalamus can produce changes in the temperature-regulating mechanism.

The anterior hypothalamus sends fibers to the **ventral tegmental gray** (14–92) of the mesencephalon in the area of the **red nucleus** (11–92) through the **anterior hypothalamic tegmental tract** (13–92), which runs through the **lateral hypothalamus.** The **posterior hypothalamic tegmental tract** (12–92) also travels to the ventral mesencephalic tegmentum. In the area dorsolateral to the **mammillary body** the two tracts run quite close together. Thus, bilateral lesions in the posterolateral area of the hypothalamus disrupt transmission in both pathways, with the result that the mechanisms for both heat loss and heat conservation are blocked. The subject is then poikilothermic.

In the mesencephalon the fibers from the hypothalamic tegmental tracts synapse with the **tegmentobulbospinal fibers** (15–92) that connect with preganglionic neurons that regulate the activity of the sweat glands and the calibration of the blood vessels. Stimulation of some areas of the ventral tegmentum have been shown to produce thermoregulatory changes. Some of the descending fibers establish connections in the **reticular formation** (16–92) of the brain stem and thus contribute to the activation of the respiratory and cardiovascular mechanisms.

The temperature-control mechanisms are also subject to influence by several cortical areas. It has been shown, for example, that stimulation of the **anterior cingulate cortex** (3–92) in humans results in shivering, piloerection, and changes in temperature. Vasoconstriction opposite the involved side results from lesions in parts of Brodmann's area 6 or parts of the internal capsule in humans. Ablation in parts of area 4 results in an increase in shivering.

SLEEP

The details of the neurology of sleep cannot be covered here. However, an attempt will be made to provide an overview of the systems involved (Fig. 93).

At least four different stages of sleep can be identified on the basis of brain waves. However, it appears that two of the stages have different neural bases. Light sleep from which the individual can be readily awakened is characterized by synchronized cortical activity. The electroencephalographic (EEG) pattern consists of spindle-shaped waves of about 11 to 16 cycles per second and by high-voltage slow waves or by both. During light sleep there is relatively little change in autonomic activity, and spinal reflexes are present. Muscle tone remains, with the result that the sleeping individual may be sitting upright and the head will remain erect.

The EEG of paradopical or REM (*rapid eye movement*) sleep is characterized by low-voltage fast waves similar to those found in the alert waking state, hence the term "paradoxical." REM sleep occurs periodically during the night and becomes more frequent as the night progresses. It is during REM sleep that dreaming occurs. If the sleeper is awakened during REM sleep, he is able to report dream content about 80 percent of the time. In REM sleep rapid eye movement occurs and may be related to the dream content. There is a depression of spinal reflexes. Autonomic changes include a drop in blood pressure, irregular breathing, and a decreased heart rate. All antigravity muscles, particularly those in the neck, lose their tone. REM sleep is deep, and it is more difficult to awaken the individual during this sleep stage.

The amount of sensory input is related to the sleep mechanism. It is, of course, common knowledge that wakefulness is generally maintained in an environment of high sensory stimulation, and one prepares for sleep by reducing external stimulation. This is due, in part, to input from the sensory pathways to the **reticular activating system (RAS),** (see Fig. 89). Lesions in several parts of the reticular activating system result in profound somnolence in the experimental subject. Sleep induced by RAS lesions can usually be disrupted by strong stimulation. The subject will awaken for a short period but will resume sleeping when the stimulation is terminated.

Sleep is also an active process. Electrical stimulation in the brain stem in the area of the **raphe nucleus** (18–93) produces slow, synchronous waves in the cerebral cortex characteristic of light sleep. Such stimulation may also result in behavioral sleep. Sleep may also be induced by direct electrical stimulation in the **intralam-**inar nuclei (4–93) of the thalamus. However, the stimulation must be of a low frequency and with a regular rhythm. High-frequency stimulation in the same area produces arousal. The behavior produced by low-frequency stimulation in the thalamus resembles that found in normal sleep. The cat circles slowly, appearing to look for a comfortable spot, and then lies down, closes its eyes, and sleeps. It can be awakened by moderate external stimulation. Sleep can also be induced in some mammals and in humans by electrical stimulation of the **caudate nucleus** (1–93). This appears to be in keeping with the general suppressor functions of that structure.

Deep pathological sleep is produced by lesions in various parts of the nervous system. Bilateral damage to the midbrain in the **periventricular gray** (11–93) in the area of the oculomotor nucleus results in deep sleep from which the patient cannot be aroused. This is one of the portions of the brain stem damaged by the disease processes in epidemic encephalitis (sleeping sickness).

If the area in the **posterior hypothalamus** (8–93) just dorsolateral and caudal to the mammillary bodies is destroyed in monkeys, the result is deep somnolence for four to eight days. The animals can be aroused but rapidly resume sleeping. These subjects may be drowsy for months. There is also considerable clinical evidence that damage to the same general area in humans by tumor or disease results in pathological somnolence.

The **preoptic region** (6–93) also seems to make a contribution to the sleep–waking mechanism. In rats only, destruction of this area results in perpetual insomnia. The subjects showed no sleep during a three-day period after which they went into a coma and died. The same lesions in other animals do not appear to have that effect. However, it has been shown that stimulation in the preoptic region produces sleep. Such stimulation also suppresses activity in the reticular activating system, and warming of the region produces drowsiness.

There remains much work to be done on the neurological basis of REM sleep. However, it does appear to be dependent on the **pontine reticular formation** (see Fig. 89). The details remain to be worked out, but the **locus ceruleus** (14–93) seems to be involved. REM sleep is eliminated if that part of the pons is lesioned. Light sleep is converted to REM sleep if the reticular formation of the pons is stimulated. The pontine area for REM sleep seems to be divided into two regions: one produces the characteristic EEG; the other sends inhibitory impulses to the muscles.

Figure 93

Neural mechanisms involved in sleep.

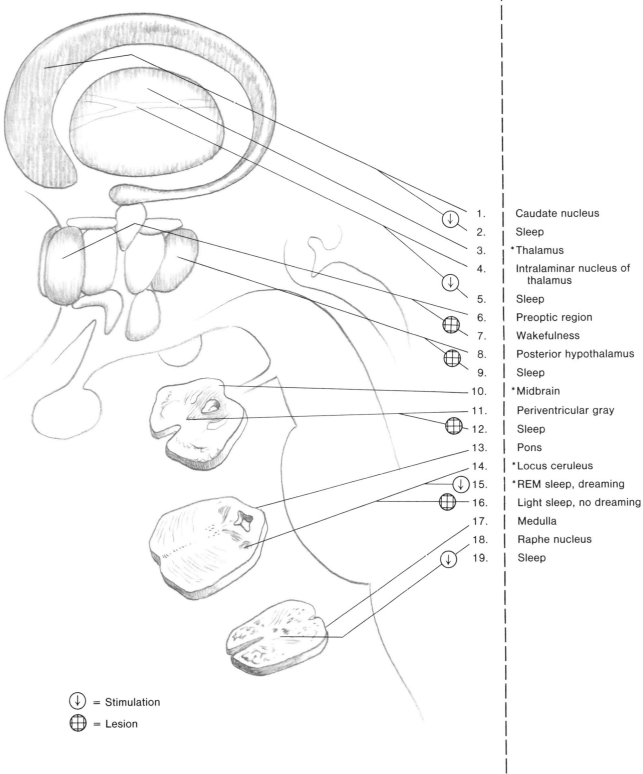

1. Caudate nucleus
2. Sleep
3. *Thalamus
4. Intralaminar nucleus of thalamus
5. Sleep
6. Preoptic region
7. Wakefulness
8. Posterior hypothalamus
9. Sleep
10. *Midbrain
11. Periventricular gray
12. Sleep
13. Pons
14. *Locus ceruleus
15. *REM sleep, dreaming
16. Light sleep, no dreaming
17. Medulla
18. Raphe nucleus
19. Sleep

↓ = Stimulation

⊕ = Lesion

WATER CONSUMPTION

Why does one begin to drink? Why does one stop drinking once started? These are simple questions with very complex answers. At the present stage of knowledge complete answers are not available. The neural and hormonal mechanisms involved in water consumption are only partially understood, and new findings are constantly appearing in the literature. However, an overview of the process can be given (Fig. 94).

Body water is lost during the process by which the kidneys clear the blood of urea. It is also lost through perspiration, the evaporation of moisture from the respiratory system, and through the feces. When the amount of water in the tissues and in the blood decreases to a given level in a normal individual, thirst is felt and drinking behavior occurs. The mechanisms for regulating water intake are remarkably precise.

Hydration level is maintained primarily by the kidneys. During the process of clearing the blood of urea, the kidneys filter about 48 gallons of fluid during a 24-hour period. Most of the fluid is automatically reabsorbed by the tubules of that organ. However, approximately 6.5 gallons are subject to control by the *antidiuretic hormone (ADH)*, also called *vassopressin* of the **posterior pituitary** (21–94). ADH increases the amount of water reabsorbed by the tubules of the kidney. If the ADH level is within the normal range, only 2 pints of fluid escape to the bladder and are subsequently excreted.

ADH is produced by secretory cells in the **supraoptic nucleus** (16–94) and the **paraventricular nucleus** (9–94) of the hypothalamus. These secretory cells send fibers down the **infundibulum** (18–94) in the **supraopticohypophysial tract** (20–94) to the posterior pituitary gland. ADH travels down the axoplasm of the producing nerves and is stored in the posterior pituitary, subject to release by neural stimulation.

Damage to the nuclei in which ADH is produced results in the disorder known as *diabetes insipidus*. Since ADH is not available to promote the reabsorption of fluid by the kidneys, all of the 6.5 gallons of fluid normally under ADH control is excreted. As a result the patient reports extreme thirst and must drink excessive amounts of water to maintain adequate levels of tissue hydration.

There is evidence that there are receptors in the brain sensitive to the effects of dehydration. One of the effects of dehydration of cells is a relative increase in the mineral concentration with a resultant increase in osmotic pressure. It seems likely, although it is not completely established, that the relevant receptors in the hypothalamus for water balance are osmoreceptors.

It is possible to stimulate precise areas of the brain with either liquid or crystalline chemicals through implanted cannulae. When a particular portion of the **lateral hypothalamus** (11–94) of water-satiated goats is stimulated with a hypertonic saline solution, the animals begin to drink. If the goats are deprived of water and then stimulated in the same area with a hypotonic saline solution, they refuse to drink. Hypertonic solutions have a higher osmotic pressure than normal, while hypotonic solutions have a lower-than-normal osmotic pressure. Injections of hypertonic saline directly into the supraoptic nucleus of the hypothalamus results in an increase in the firing rate of individual cells.

The neural systems for water and food consumption overlap in the lateral hypothalamus. However it is most likely that different sets of cells are involved in each system. Satiated animals will show increased drinking if cholinergic compounds are introduced directly into the lateral hypothalamus. Adrenergic compounds applied to the same area result in eating behavior by a satiated subject. Electrical stimulation of this general hypothalamic area results in the release of ADH into the blood stream. Thus, water is conserved because there is increased reabsorption in the kidney tubules.

Lesions in the lateral hypothalamic portion of the neural system for water consumption result in the elimination of drinking behavior (*adipsia*). These lesions also result in *aphagia,* or loss of eating behavior. Some recovery of function does occur. However, the drinking behavior that becomes manifest is only that associated with the eating of dry food. Drinking in response to activation of the usual dehydration mechanisms, such as changes in osmotic pressure, does not recover.

Water consumption, like other complex motivated behaviors, is not controlled by a particular center. A system is involved that runs through several portions of the brain. Many investigators are attacking this problem, but the details of the neural system for the maintenance of water balance have not yet been worked out. It has been shown, however, that a number of structures in the limbic system are involved. Drinking will occur in satiated animals if carbachol (a cholinergic substance) is directly applied to portions of the **hippocampus** (7–94), the **septal region** (3–94), the **anterior nucleus of the thalamus** (5–94), and the **cingulate gyrus** (1–94).

Figure 94

Neural mechanisms involved in the control of water consumption.

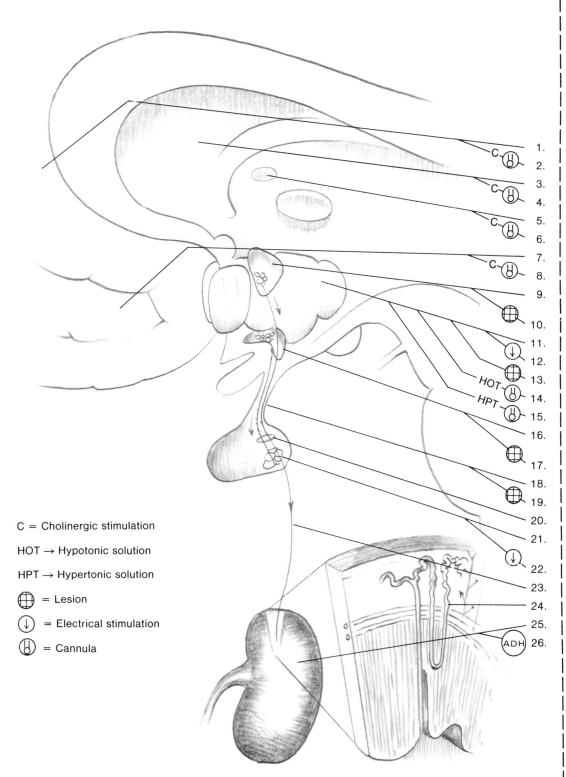

C = Cholinergic stimulation

HOT → Hypotonic solution

HPT → Hypertonic solution

⊕ = Lesion

↓ = Electrical stimulation

⊎ = Cannula

1. Cingulate gyrus
2. Drinking
3. Septal region
4. Drinking
5. Anterior nucleus of thalamus
6. Drinking
7. *Hippocampus
8. Drinking
9. Paraventricular nucleus, produces ADH
10. Diabetes insipidus
11. Lateral hypothalamus
12. *ADH release
13. Adipsia
14. Adipsia in thirsty animal
15. Drinking in satiated animal
16. Supraoptic nucleus, produces ADH
17. Diabetes insipidus
18. Infundibulum
19. Diabetes insipidus
20. Supraopticohypophysial tract
21. Posterior *pituitary, stores ADH
22. *ADH release
23. *ADH in blood stream
24. Kidney tubules
25. Kidney
26. Promotes water reabsorption

FOOD CONSUMPTION

The neural mechanisms controlling the consumption of nutritive substances are even more complex than those responsible for water consumption (Fig. 95). Food consumption is influenced by a variety of factors, including general and specific metabolic needs, taste, odor, and foods available. Learned preferences also play a significant role and may override some of the more basic regulatory mechanisms. Our concern in this section will be primarily with the central neural mechanisms involved in the control of eating behavior.

At one time the evidence seemed to indicate rather clearly that mammals began or ceased eating as a result of the functioning of a feeding center in the **lateral hypothalamus (LH)** (6–95) and a satiety center in the area of the **ventromedial nucleus of the hypothalamus (VMH)** (16–95). More recent research leads to the conclusion that the idea of controlling centers is too simple an explanation for consumatory behavior, just as it is for other complex motivated behaviors. The lateral and ventromedial hypothalamus play an important, but not exclusive, role in the regulation of food intake. Let us examine the evidence first and then look at some of the more recent research.

It has been known for many years that lesions in the area of the ventromedial hypothalamus result in an animal specimen that will eat excessively and become obese. This syndrome is called *hypothalamic hyperphagia (HH)*. The symptoms associated with HH are complex. Shortly after surgery the animal goes through a dynamic phase in which it consumes large amounts of food and shows a rapid weight gain. It eats longer rather than more frequent meals. In the second (static) phase, after a state of considerable obesity is reached, the animal eats a normal amount but the excess weight is maintained. HH animals do not have an abnormal metabolism; they become fat because they eat excessive amounts. They are more lethargic than normal animals and are finicky about their food. If the food is adulterated with bitter-tasting quinine, they will eat less than a normal animal. However, if the food is adulterated with saccharin, which has no nutritive value, they will eat more than a normal subject. These animals will eat more than normal subjects, but paradoxically they are less hungry if hunger is defined as the amount of work that will be done to get food. These subjects will not press a bar as often as normal subjects to get food, and they will not take as strong a shock to get it. The excessive eating of HH animals leads to the conclusion that ventromedial lesions interfere with a "satiety center." Thus, the animals have a damaged stop-eating mechanism.

Other experiments tend to support this position. For instance, stimulation of the VMH either electrically or with hypertonic saline inhibits food intake in a deprived animal. However, such stimulation has also been shown to be aversive, which may be the cause of the cessation of eating. Procaine is an anesthetic that temporarily blocks neural functioning. If it is applied directly to the VMH through a cannula, a satiated animal will begin to eat. It has also been suggested that the VMH contains glucose receptors. During hyperglycemia there is an increase in the rate at which the neurons of the VMH fire. That rate decreases during hypoglycemia.

Also, bilateral lesions in the so-called feeding center, located in the lateral hypothalamus, produce the lateral hypothalamic syndrome. The animal refuses all food and behaves as though food intake were aversive. This refusal to eat, *aphagia,* is accompanied by a loss of water consumption, *adipsia.* If the animal is not force fed, it may starve to death in the presence of food. Animals with LH syndrome gradually recover from the aphagia (although generally not from the adipsia). Taste is an important variable in that they will eat preferred foods such as chocolate but reject all other food. If the subjects are well cared for, they will progress from eating preferred wet foods to eating enough normal food to maintain their weight, although at a reduced level.

Also, electrical stimulation of the LH will cause a satiated animal to eat and to continue doing so for some time after the cessation of the stimulation. The animal will also perform learned responses in order to obtain food, which indicates that reflex eating is not the explanation. The direct application of adrenergic substances to the LH also increases eating, while the application of procaine will block it. The rate at which the neurons in the LH fire is increased during hypoglycemia and is reduced during the state of hyperglycemia.

The evidence of these additional experiments all leads to the conclusion that the VMH is important in satiation and that the LH is important in the initiation of food intake. One need not conclude, however, that these are the "centers" that control nutritive consumatory behavior. Recent studies seem to indicate it is more likely that these are the areas through which neural systems for food consumption run. A very careful study of VMH le-

(Continued)

Figure 95

Neural mechanisms involved in the control of food consumption.

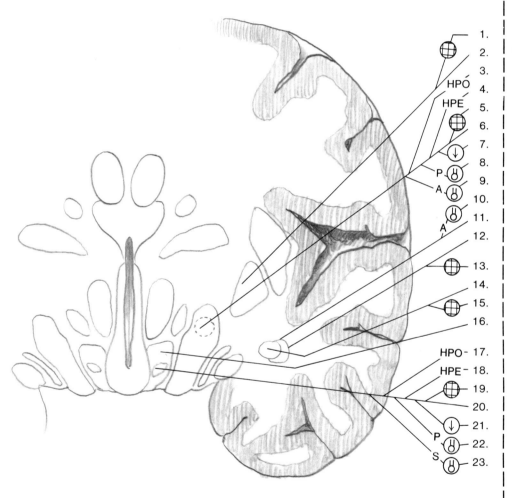

1. Aphagia
2. *Globus pallidus
3. Decrease in impulse rate
4. Increase in impulse rate
5. Aphagia and adipsia
6. Lateral hypothalamus
7. Eating
8. Stop eating
9. Eating
10. Eating
11. Amygdala
12. Amygdala, lateral and basal nuclei
13. Hyperphagia
14. Amygdala, anterior nucleus
15. Aphagia
16. Ventromedial nucleus, hypothalamus
17. Decrease in impulse rate
18. Increase in impulse rate
19. Hyperphagia and obesity
20. Ventral noradrenergic bundle
21. Stop eating
22. Eating
23. Stop eating

⊕ = Lesion   HPE = Hyperglycemia   A = Adrenergic compound

↓ = Electrical stimulation   HPO = Hypoglycemia   S = Saline

⊗ = Cannula   P = Procaine

213

sions showed that the syndrome of hypothalamic hyperphagia did not occur with VMH damage unless the tract known as the **ventral noradrenergic bundle** (20–95) was destroyed. It was further shown that lesions need not be bilaterally symmetrical. A VMH lesion that destroyed that pathway on one side effectively produced HH if it was combined with a lesion on the other side that cut the tract at the level of the midbrain.

Other areas of the brain are also important in satiation, which implies that a rather complex system is involved. Lesions in the basal and lateral portions of the **amygdala** (11–95) produce hyperphagia, as do lesions in some parts of the frontal lobe.

Aphagia can also be produced by the lesioning of the anterior amygdala and of some tracts that send fibers through the lateral hypothalamus. Interestingly some of these structures are associated with the extrapyramidal system, including the **globus pallidus** (2–95) and the **substantia nigra** (14–79). Aphagia is also caused by lesions in the trigeminal lemniscus, which sends fibers through the LH. Eating can be induced by the electrical stimulation of the anterior cingulate gyrus or the midline thalamus. It can also be induced by the stimulation of the amygdala with norepinephrine.

It is still far from clear how all of these different areas of the brain interact to control food consumption. It is clear, however, that the mechanism is extremely complex and that a simple two-center theory is not acceptable.

SEXUAL BEHAVIOR

Sexual behavior is a function of a complex interaction of neural, hormonal, experiential, and situational components. There are, of course, differences in the physiological substrates of that behavior in males and females, and there are rather wide differences across species. Differences in sexual behavior among species are, in general, much greater than differences in more-basic mechanisms, such as temperature regulation or respiration control.

The females of some animal species manifest a periodic sexual receptivity known as the estrus cycle. Estrogen is a more important determinant of sexual responsiveness in those species than it is in species such as monkeys and humans, which do not show estrus. In rats and cats ovariectomy, which essentially eliminates estrogen production, also inhibits the sexual receptivity of those animals. The same operation in female monkeys or humans has relatively little effect.

The sexual behavior of males, including humans, is more hormonally dependent. Although some males continue to be capable of sexual performance after castration, for most, libido is significantly reduced by the operation. Bremer, who studied 215 legally castrated men in Norway, reported that without exception their sexual behavior was reduced or abolished.

The hypothalamus is an important part of the neural system for mating behavior and it has two types of influence (Fig. 96). The area posterior to the **infundibulum** (21–96) controls the secretion of the gonadatrophic hormones from the **anterior pituitary** (22–96). Thus, lesions in the **medial and lateral mammillary nuclei** (18–96) in male rats result in gonadal atrophy and a reduction in sexual behavior. This behavior will be restored by the administration of testosterone. Male rats electrically stimulated in the **posterior hypothalamus** (16–96) will readily copulate with an estrus female. They do not make inappropriate responses in the absence of the female, but will work at a learned response to get to a female. Mammillary and premammillary lesions in the female cat will also eliminate sexual receptivity that can be restored by the administration of estrogen. Thus, the caudal hypothalamus influences sex responses indirectly through the hormonal system.

There are also areas of the hypothalamus that have a direct effect on and are receptive to the influence of the hormones. Large lesions in the **preoptic nuclei** (9–96) of the male rat eliminate sexual response, and the behavior

cannot be restored by the peripheral administration of androgens. Electrical stimulation of the **lateral preoptic area** (9–96) of the male rat, however, results in a dramatic increase in sexual performance. One animal achieved 45 ejaculations in a prolonged seven-hour satiation test. Direct application of testosterone to the same area results in indiscriminate mounting of any available target by the rat, and will reestablish the sexual capacity of rats that have been castrated.

Lesions in the preoptic area of female rats facilitates the precoital posturing called lordosis. The subjects are receptive for longer periods and less estrogen is required to make them receptive. However, lesions in the **anterior nuclei** (13–96) just caudal to the preoptic area reduce sexual receptivity in female rats and cats. This receptivity is not restored by the peripheral administration of estrogen. The direct implantation of tiny amounts of estrogen in the anterior nuclei of the hypothalamus in female rats and cats will restore sex behavior that has been eliminated by ovariectomy. Estrogen implants in other brain areas do not have that effect.

Reflex mechanisms for some portions of the sexual response for both males and females are located in the spinal cord. Penile erection and ejaculation can be elicited by genital stimulation in dogs with the spinal cord transected. This reflex can also be induced in humans with severed cords. They do not, of course, experience an orgasm. Female rats with the cord cut below the brain will also show sex movements and postures if the genital region is stimulated. Normally these spinal reflexes are inhibited by impulses from the cortex.

The limbic system is also an important part of the neural system for sexual behavior and receptivity. However, the details of the relationship between the limbic system and the hypothalamus remain to be worked out, and the connections between the various components within the limbic system are not well understood.

A variety of lesions in the **temporal lobe** (3–96) may have dramatic effects on the sexual responsiveness of the subject. Bilateral temporal lobectomy results in the complex Kluver-Bucy syndrome, which includes loss of aggressiveness, overeating, and hypersexuality. A monkey with these lesions will attempt to mount anything from a female monkey to a piece of furniture, and at least one human patient has shown the essential components of Kluver-Bucy syndrome after bilateral temporal lobe lesions. Male cats and monkeys with lesions of the

(Continued)

Figure 96

Neural mechanisms involved in sexual behavior.

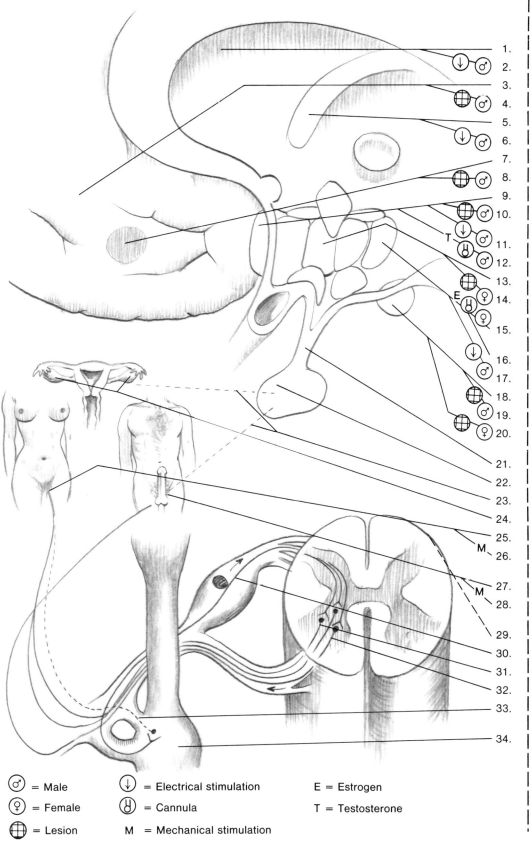

1. Septum
2. Sexual ideation
3. Temporal lobe
4. Hypersexuality
5. Fornix
6. Sexual ideation
7. Amygdala
8. Hypersexuality
9. Lateral preoptic region
10. Decreased sexual behavior, not restored by androgens
11. Increased sexual behavior
12. Increased sexual behavior
13. Anterior hypothalamic region
14. Decreased sexual behavior, not restored by estrogen
15. Sexual behavior restored in spayed females
16. Posterior hypothalamus
17. Increase in sexual behavior
18. Mammillary body
19. Gonadal atrophy
20. Decreased sexual behavior, restored by estrogen
21. Infundibulum
22. Anterior *pituitary
23. Gonadatrophic hormones
24. Ovaries
25. Female genitals
26. Sexual postures and movements (reflexive)
27. Male genitals
28. Erection and ejaculation (reflexive)
29. *Spinal cord
30. Sensory neuron
31. Parasympathetic fiber
32. Somatic nerve fiber
33. Postganglionic sympathetic fiber
34. Sympathetic ganglion

⚥ = Male   ⬇ = Electrical stimulation   E = Estrogen

♀ = Female   cannula = Cannula   T = Testosterone

⊕ = Lesion   M = Mechanical stimulation

217

amygdala (7–96) also show extreme hypersexuality. Sexual dysfunctions are also found in humans who have irritative lesions in the temporal lobe; the disorder may be either an increase or a decrease in sexuality. In some cases where neurosurgeons have used stereotaxic lesions in the temporal lobe to control intractable epilepsy, the operation has been reported to result in hypersexuality in male patients.

Direct electrical stimulation of several areas in the human brain have been reported to result in erotic ideation. The male patient may be conversing about a serious subject such as a member of the family who is ill and immediately after stimulation offer to tell the physician an off-color story or indicate that he plans to seduce the waitress at the corner bar. The patient is, of course, not aware that his brain has been stimulated. This finding has been reported during stimulation in the **septum** (1–96), in the **fornix** (5–96), and in an unspecified portion of the temporal lobe. Essentially nothing more is known about this phenomenon.

AGGRESSIVE BEHAVIOR

There is considerable confusion about the neurological bases of aggressive behavior because it is frequently not recognized that aggression is not a unitary phenomenon, but that there are, in fact, many kinds of aggressive behavior. I have suggested elsewhere[2] that there are at least seven different kinds of aggression, including predatory, intermale, fear-induced, irritable, maternal, instrumental, and sex-related. It is, of course, well beyond the scope of this discussion to attempt to cover what is known about the neurology of all of these different kinds of aggression. We will therefore concentrate on irritable aggression, which in humans is accompanied by the subjective state of anger. (See Fig. 97.)

> Relatively little is known about the *detailed* neuro-anatomy of the neural systems for aggressive behavior in human beings. However, the general outline is clear. The weight of the evidence seems to indicate that for all of their encephalization humans have not escaped the biological determinants of hostility. They have neural systems for aggressive feelings and behavior, and when these systems are activated, by whatever means, they have aggressive feelings and may act aggressively. When the activity in the neural systems is blocked, either by surgical interruption or by suppression from inhibitory systems, the tendency to hostility is reduced.[3]

Most of our understanding of the neuroanatomical bases of irritable aggression is based on animal experiments, and one must be cautious about assuming that those data are directly relevant to human beings.

The **ventromedial hypothalamus (VMH)** (27–97) is clearly involved in irritable aggression. Direct electrical stimulation in this area will cause an animal to attack either the experimenter or another animal. If the sites from which irritable attack can be elicited are lesioned, the subsequent neural degeneration can be traced. It is essentially confined to the **periventricular system** (26–97) that connects the medial hypothalamus and the **central gray of the midbrain** (30–97). Stimulation of the central gray produces attack behavior that may be related to pain.

Lesions in the ventromedial hypothalamus will also result in an increase in aggressive behavior and, in humans, in feelings of hostility. This has been reported in cases of VMH tumors. It is not yet clear whether this is due to the irritative nature of the lesions or whether a suppressor system has been interrupted. Lesions in the **posterior hypothalamus** (24–97) have been reported to reduce pathological hostility in humans.

The **cingulum** (14–97) is involved in aggressive behavior, but there are species differences. Either stimulation or lesions (again the effect may be due to irritation) will increase irritability in cats. Cingulate lesions, however, reduce aggression in monkeys and in humans.

The role of the **amygdala** (3–97) in aggression is extremely complex. It is involved in the neural circuitry of at least three different kinds of aggressive behavior. Early studies showed that total bilateral amygdalectomy produced a dramatic reduction in irritable aggression. Amygdalectomized cats do not aggress even when suspended by their tails or when they are generally roughed up. More precise experiments indicate that the **central nucleus of the amygdala** (6–97) may function to inhibit irritable aggression. Lesions of the central nucleus result in an increase in irritability in cats, whereas stimulation produces a blocking of the "rage" response induced by hypothalamic stimulation. Stimulation of the **medial nucleus of the amygdala** (9–97) causes an increase in aggressiveness in dogs and cats, whereas lesions in that area tend to reduce it. It has also been shown that direct electrical stimulation of the amygdala of humans results in feelings of hostility and verbal threats and gestures.

Portions of the **hippocampus** (12–97) are probably involved in irritable aggression, but it is not clear how this structure fits into the overall pattern. It has been shown in cats that hippocampal lesions result in an increase in aggressiveness toward the experimenter and toward other cats. Hippocampal-lesioned rats also become more aggressive in some situations.

The **septum** (17–97) is a complex structure involving several nuclei that may be concerned with quite different systems. Lesion experiments frequently involve considerable destruction of this structure and are therefore difficult to interpret. However, in some species at least, the septum does seem to be involved in affective, or irritable, aggression. Rats with septal lesions show a temporary increase in irritability. It has also been demonstrated that direct septal stimulation in humans can inhibit strong, ongoing irritable aggressive behavior.

---

[2] K. E. Moyer, *The Psychobiology of Aggression* (New York: Harper & Row, 1976).
[3] Ibid., p. 57.

(Continued)

Figure 97

Neural mechanisms involved in irritable aggression.

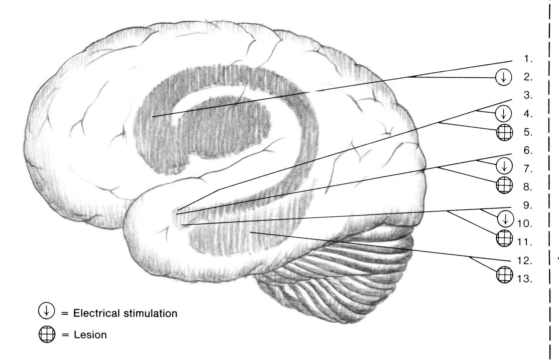

1. Caudate nucleus
2. Aggression inhibition
3. Amygdala
4. Aggression increase
5. Aggression decrease
6. Central nucleus of amygdala
7. Aggression decrease
8. Aggression increase
9. Medial nucleus of amygdala
10. Aggression increase
11. Aggression decrease
12. *Hippocampus
13. Aggression increase

⊘ = Electrical stimulation

⊕ = Lesion

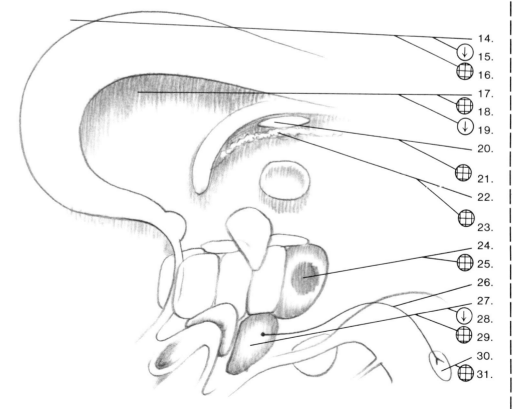

14. Cingulate gyrus
15. Aggression increase
16. Aggression decrease
17. Septum
18. Aggression increase
19. Aggression inhibition
20. Dorsomedial nucleus of thalamus
21. Aggression decrease
22. Intralaminar nucleus of thalamus
23. Aggression decrease
24. Posterior hypothalamus
25. Aggression decrease
26. Periventricular system
27. Ventromedial hypothalamus
28. Aggression increase
29. Aggression increase
30. *Central gray
31. Aggression increase

Several investigators have reported that thalamic lesions exert a significant inhibitory effect on pathological aggressive behavior. Aggressiveness and assaultive behavior subside after lesions of the **dorsomedial nuclei** (20–97); lesions in the **intralaminar nuclei** (22–97) are particularly effective in reducing aggression and destructive behavior.

The **caudate nucleus** (1–97) appears to have an inhibitory effect on irritable aggression, at least in monkeys. Direct stimulation of the caudate nucleus of a boss monkey reduces its aggressiveness toward other colony members with the result that the boss loses his dominant status. When stimulation is terminated the aggressiveness returns.

PSYCHOSURGERY

For a number of years there has been a controversy about the efficacy and the ethics of psychosurgical procedures. It is not the purpose of this section to contribute to that controversy but to introduce students briefly to the major psychosurgical techniques (Fig. 98).

Modern psychosurgery was initiated by Antonio Egas Moniz in 1935 in the form of the prefrontal lobotomy. A major factor contributing to his decision to develop the operation was a report of an experiment on chimpanzees by Jacobson in which it was shown that intense emotionality connected with the experimental situation was eliminated by bilateral frontal lobectomy. In his initial report on 20 patients, Moniz indicated that 14 had either recovered or improved and that agitated, depressed patients were the best candidates for the operation.

It is estimated that between 1935 and the early 1950s as many as forty thousand prefrontal lobotomies were performed in the United States alone. The number of operations was drastically reduced after the advent of psychotrophic drugs a few years after the Second World War. Prefrontal lobotomies are now done with relative infrequency.

There have been a large number of theoretical justifications for the lobotomy procedure. Understandably these theories have been expressed in rather vague terms. The aim of the surgery has generally been to disrupt the connections between the frontal lobes, in which emotional reactions are presumably integrated, and the thalamus, which receives the principal projections from the lower areas active during emotional experiences. Thus, the prime target has been the thalamo-frontal radiation. A number of techniques have been developed to achieve this goal. One of the most frequently used was the so-called **precision lobotomy** (1–98) of Freeman and Watts. Burr holes were drilled in the skull, and a blunt surgical tool was inserted through the holes and swung up and down in an arc. The operation was *minimal, standard,* or *radical* depending on how much of the frontal lobes were separated from the rest of the brain. The operation was generally done under local anesthesia with the patient conscious.

Freeman greatly simplified the lobotomy procedure when he introduced the **transorbital lobotomy** (2–98) to the United States from Italy. In this technique a sharp probe was introduced under the patient's eyelid and driven through the orbit with a light tap from a hammer. Once introduced into the brain the instrument could be swung from side to side, cutting frontal-lobe connections. Freeman considered the transorbital lobotomy to be a "minor operation"; patients could be released from the hospital within two days, and some of these lobotomies were even done in the physician's office.

In 1949 Spiegel and Wycis introduced the **thalamotomy** (7–98). In this procedure a stereotaxic instrument was used to place an electrode in the dorsomedial nucleus of the thalamus. To guide the electrode placement they used X rays that showed the pineal gland and the ventricles. The electrode was then used to make electrolytic lesions, which rendered many of the thalamofrontal radiations nonfunctional. Thalamotomy was presumed to have the same effects as lobotomy with less surgical trauma and fewer side effects.

Finally, a prefrontal technique called **topectomy** (3–98) was introduced. This operation involved lesioning or undercutting three specific Brodmann's areas, 9, 10, and 46.

The evaluation of the effectiveness of the various lobotomy procedures will be left to other sources. However, it should be pointed out that in spite of a variety of unfortunate side effects, a large number of hopelessly ill individuals were helped by the procedure. There is little evidence to support the Sunday-supplement portrayal of the lobotomized patient as little more than a living vegetable. Currently, however, there is little justification for using prefrontal lobotomies. Psychopharmacological approaches are more effective and less hazardous. The operation is still occasionally used for the control of intractable pain, but it seems clear that other methods, such as chordotomy, should be tried first and that lobotomy should be an absolute last resort.

Recently more-limited and more-precise psychosurgical procedures have been introduced, most frequently involving stereotaxic procedures. For instance, lesions in the anterior portion of the cingulate gyrus (5–98) were introduced as a psychosurgical procedure for the control of anxiety, tension, depression, and agitation. This **cingulotomy** was initiated on the basis of animal studies showing the involvement of the cingulum in emotional responses. (It should be pointed out, however, that these animal experiments were, at best, equivocal.) More recently **posterior cingulotomy** (6–98) has been used in an attempt to control chronic aggressiveness, hostility, and antisocial tendencies.

As the last few sections of this book have shown, the

(Continued)

Figure 98

Psychosurgical techniques.

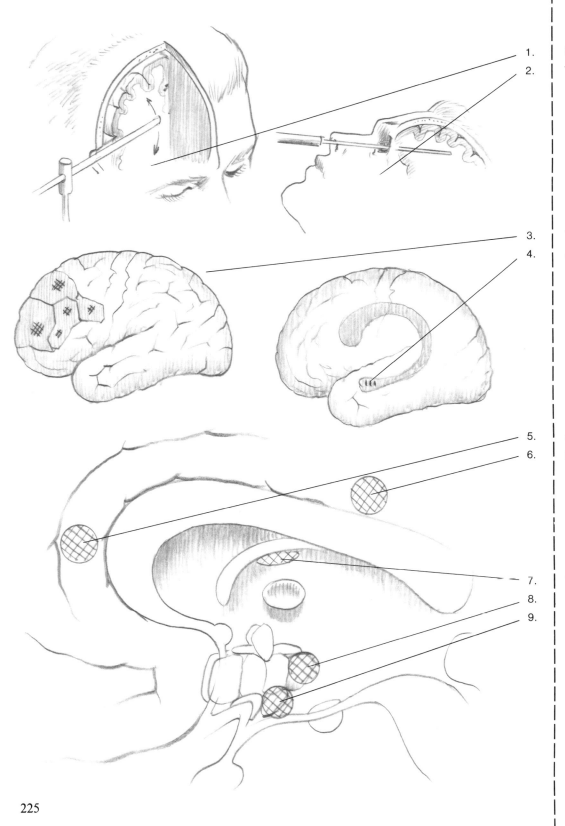

1. Precision lobotomy
2. Transorbital lobotomy

3. Topectomy
4. Amygdalotomy

5. Anterior cingulotomy
6. Posterior cingulotomy

7. Thalamotomy
8. Posterior hypothalamotomy
9. Ventromedial hypothalamotomy

functions of the hypothalamus are many, varied, vital, and overlapping. No portion of that complex structure serves a single function. One would suspect that hypothalamic lesions would be made only with great trepidation. However, Dr. Keiji Sano at the University of Tokyo has introduced an operation that he refers to as *sedative surgery.* He lesions the **posterior hypothalamus** (8–98) in order to control patients showing intractable violent behavior. He reports that they become markedly calm, passive, and tractable with a decrease in spontaneity. Also, Roeder and his colleagues in Germany have been so bold as to destroy the ventromedial nucleus of the hypothalamus (9–98) in order to reduce pedophilic homosexuality. Although they have reported success in some patients, further evaluation is obviously necessary.

A number of surgeons now use **amygdalotomy** (4–98) in an attempt to control excessive aggressive behavior. Lesions from 8 to 10 millimeters in diameter are produced by thermocoagulation, cryosurgery, or the injection of oil to which Lipiodol is added. Destructiveness, hostility, and aggression toward others are the behavior symptoms most frequently improved by the operation. One investigator has concluded that stereotaxic amygdalotomy is a safe and relatively easy procedure that can be used to treat a select group of patients who would be otherwise untreatable.

APPENDIX of ALTERNATE TERMS

In this appendix the alternate terms are listed. The preferred term, that is, the one most frequently used in this book, is printed in **boldface** type, whereas those used less frequently are listed in *italics*.

accessory nerve [XI] = *spinal accessory nerve*
acoustic nerve [VIII] = *stato-acoustic nerve* = *vestibulo-cochlear nerve*
ala cinerea = **trigonum vagi**
Ammon's horn = **hippocampus** = *cornu ammonis*
ansiform lobule = **crus I and crus II**
anterior crescentic lobule = **anterior semilunar lobule** = *anterior quadrangular lobule*
anterior quadrangular lobule = **anterior semilunar lobule** = *anterior crescentic lobule*
anterior semilunar lobule = *anterior crescentic lobule* = *anterior quadrangular lobule*
antidiuretic hormone (ADH) = *vassopressin*
aqueduct of Sylvius = **cerebral aqueduct**
arachnoid granulations = *pacchionian bodies*
arcuate nucleus of the thalamus = **ventral posterior nucleus of the thalamus** = *semilunar nucleus of the thalamus*
baroreceptors = *barostats*
barostats = **baroreceptors**
basilar pons = **pars ventralis**
basis pedunculi = **cerebral peduncle** = *crus cerebri*
Betz cell = **pyramidal cell**
biventer lobule = *dorsal paraflocculus*
brachium conjunctivum = **superior cerebellar peduncle**
brachium pontis = **middle cerebellar peduncle**
caput = *head*
cauda equina = *horse's tail*
central fissure = *fissure of Rolando*
central gray stratum = *periaqueductal gray*
cerebellomedullary cistern = *cisterna magna*
cerebral aqueduct = *aqueduct of Sylvius*
cerebral peduncle = *basis pedunculi* = *crus cerebri*
circular fissure = **limiting fissure**
cisterna magna = *cerebellomedullary cistern*
clava = **gracilis tubercle**
cold-blooded = *poikilothermic*
cornu ammonis = **hippocampus** = *Ammon's horn*
corpora quadrigemina = **tectum** = *quadrigeminal bodies* = *quadrigeminal plate*
crossed pyramidal tract = **lateral cerebrospinal fasciculus**
crus cerebri = **cerebral peduncle** = *basis pedunculi*

crus I = **superior semilunar lobule**
crus II = **inferior semilunar lobule**
crus I and crus II = *ansiform lobule*
decussation of Forel = **ventral tegmental decussation**
decussation of the medial lemniscus = *sensory decussation*
Deiters' nucleus = **lateral vestibular nucleus**
dentate fascia = **dentate gyrus**
dentate gyrus = *dentate fascia*
direct pyramidal tract = **ventral cerebrospinal fasciculus**
dorsal column = **dorsal horn**
dorsal horn = *dorsal column*
dorsal paraflocculus = **biventer lobule**
dorsal tegmental decussation = *fountain decussation of Meynert*
dorsal thalamus = **thalamus**
dorsolateral fasciculus = *tract of Lissauer*
epiphysis cerebri = **pineal body**
external granular layer II = *layer of small pyramidal cells*
fasciculus = **tract**
fasciculus arcuatus = **superior longitudinal fasciculus**
fasciculus cuneatus = *tract of Burdach*
fasciculus gracilis = *tract of Goll*
fastigial nucleus = *nucleus tecti*
fissure = **sulcus**
fissure of Rolando = **central fissure**
fissure of Sylvius = **lateral fissure**
foramen of Luschka = **lateral aperture**
foramen of Magendie = **medial aperture of the fourth ventricle**
foramen of Monro = **interventricular foramen**
forebrain = *prosencephalon*
fornicate lobe = **limbic lobe**
fountain decussation of Meynert = **dorsal tegmental decussation**
fusiform cell = **polymorphous cell**
fusiform gyrus = **occipitotemporal gyrus**
ganglionic layer = **internal pyramidal layer V**
ganglion of Scarpa = **vestibular ganglion**
Gasserian ganglion = **semilunar ganglion**
globus pallidus = *pallidum*
gracilis tubercle = *clava*
granule cell = *stellate cell*
ground bundles = **proper fasciculi**
head = *caput*
hindbrain = *rhombencephalon*
hippocampal gyrus = *parahippocampal gyrus*

227

hippocampus = *Ammon's horn* = *cornu ammonis*
homiothermic = **warm-blooded**
horse's tail = **cauda equina**
hypophysis = **pituitary**
indusium griseum = **supracallosal gyrus**
inferior cerebellar peduncle = *restiform body*
inferior colliculi = *postgemina*
inferior semilunar lobule = *crus II*
insula = *island of Reil*
intercalated nucleus = **intermediate mammillary nucleus**
intermediate mammillary nucleus = *intercalated nucleus*
intermediate mass = *massa intermedia* = *interthalamic adhesion*
intermediate nerve = *nervus intermedius*
internal pyramidal layer V = *ganglionic layer*
interpeduncular fossa = *posterior perforated substance*
interthalamic adhesion = **intermediate mass** = *massa intermedia*
interventricular foramen = *foramen of Monro*
island of Reil = **insula**
lateral aperture = *foramen of Luschka*
lateral cerebrospinal fasciculus = *crossed pyramidal tract*
lateral column = **lateral horn**
lateral fissure = *fissure of Sylvius*
lateral horn = *lateral column*
lateral vestibular nucleus = *Deiters' nucleus*
layer of fusiform cells = **polymorphic cell layer VI**
layer of small pyramidal cells = **external granular layer II**
lenticular nucleus = *lentiform nucleus*
lentiform nucleus = **lenticular nucleus**
limbic lobe = *fornicate lobe*
limiting fissure = *circular fissure*
lobulus simplex = **posterior semilunar lobule** = *simple lobe* = *posterior quadrangular lobule*
massa intermedia = **intermediate mass** = *interthalamic adhesion*
medial aperture of fourth ventricle = *foramen of Magendie*
medial vestibular nucleus = *Schwalbe's nucleus*
medulla spinalis = **spinal cord**
mesencephalon = **midbrain**
midbrain = *mesencephalon*
middle cerebellar peduncle = *brachium pontis*
molecular layer I = *plexiform layer*
myotatic reflex = **stretch reflex**
nervus intermedius = **intermediate nerve**
neuropituitary = **posterior lobe of the pituitary**

nuclei pontis = **pontine nuclei**
nucleus of Bechterew = **superior vestibular nucleus**
nucleus locus ceruleus = *pigmented nucleus*
nucleus tecti = **fastigial nucleus**
occipitotemporal gyrus = **fusiform gyrus**
pacchionian bodies = **arachnoid granulations**
pallidum = **globus pallidus**
paradoxical sleep = *REM sleep* = *rapid eye movement sleep*
parahippocampal gyrus = **hippocampal gyrus**
paraterminal body = **subcallosal gyrus**
pars dorsalis = *pontine tegmentum*
pars ventralis = *basilar pons*
periaqueductal gray = **central gray stratum**
perpendicular fasciculus = *vertical occipital fasciculus*
pigmented nucleus = **nucleus locus ceruleus**
pineal body = *epiphysis cerebri*
pituitary = *hypophysis*
plexiform layer = **molecular layer I**
poikilothermic = **cold-blooded**
polymorphic cell layer VI = *layer of fusiform cells*
polymorphous cell = *fusiform cell*
pontine nuclei = *nuclei pontis*
pontine tegmentum = **pars dorsalis**
postclival fissure = **posterior superior fissure**
posterior lobe of the pituitary = *neuropituitary*
posterior perforated substance = **interpeduncular fossa**
posterior quadrangular lobule = **posterior semilunar lobule** = *simple lobe* = *lobulus simplex*
posterior semilunar lobule = *simple lobe* = *lobulus simplex* = *posterior quadrangular lobule*
posterior superior fissure = *postclival fissure*
posterolateral fissure = *postnodular fissure*
postgemina = **inferior colliculi**
postnodular fissure = **posterolateral fissure**
preclival fissure = **primary fissure**
pregemina = **superior colliculi**
primary fissure = *preclival fissure*
proper fasciculi = *ground bundles*
prosencephalon = **forebrain**
pterygopalatine ganglion = **sphenopalatine ganglion**
pyramidal cell = *Betz cell*
quadrigeminal bodies = **tectum** = *corpora quadrigemina* = *quadrigeminal plate*
quadrigeminal plate = **tectum** = *corpora quadrigemina* = *quadrigeminal bodies*
rapid eye movement sleep = **paradoxical sleep** = *REM sleep*

REM sleep = **paradoxical sleep** = *rapid eye movement sleep*

restiform body = **inferior cerebellar peduncle**

reticular formation = **reticular substance**

reticular substance = *reticular formation*

rhombencephalon = **hindbrain**

Schwalbe's nucleus = **medial vestibular nucleus**

semilunar ganglion = *Gasserian ganglion*

semilunar nucleus of the thalamus = **ventral posterior nucleus of the thalamus** = *arcuate nucleus of the thalamus*

sensory decussation = **decussation of the medial lemniscus**

simple lobe = **posterior semilunar lobule** = *lobulus simplex* = *posterior quadrangular lobule*

sphenopalatine ganglion = *pterygopalatine ganglion*

spinal accessory nerve = **accessory nerve [XI]**

spinal cord = *medulla spinalis*

stato-acoustic nerve = *vestibulocochlear nerve* = **acoustic nerve [VIII]**

stellate cell = **granule cell**

stretch reflex = *myotatic reflex*

subcallosal gyrus = *paraterminal body*

subthalamus = *ventral thalamus*

sulcus = **fissure**

cerebellar peduncle = *brachium conjunctivum*

superior colliculi = *pregemina*

superior longitudinal fasciculus = *fasciculus arcuatus*

superior semilunar lobule = *crus I*

superior vestibular nucleus = *nucleus of Bechterew*

supracallosal gyrus = *indusium griseum*

sympathetic nervous system = *thoracolumbar system*

tectum = *quadrigeminal plate* = *corpora quadrigemina* = *quadrigeminal bodies*

thalamus = *dorsal thalamus*

thoracolumbar system = **sympathetic nervous system**

tonsil = *ventral paraflocculus*

tract = **fasciculus**

tract of Burdach = **fasciculus cuneatus**

tract of Goll = **fasciculus gracilis**

tract of Lissauer = **dorsolateral fasciculus**

trigonum vagi = *ala cinerea*

vassopressin = **antidiuretic hormone (ADH)**

ventral cerebrospinal fasciculus = *direct pyramidal tract*

ventral column = **ventral horn**

ventral horn = *ventral column*

ventral paraflocculus = **tonsil**

ventral posterior nucleus of the thalamus = *semilunar nucleus of the thalamus* = *arcuate nucleus of the thalamus*

ventral tegmental decussation = *decussation of Forel*

ventral thalamus = **subthalamus**

vertical occipital fasciculus = **perpendicular fasciculus**

vestibular ganglion = *ganglion of Scarpa*

vestibulocochlear nerve = **acoustic nerve [VIII]** = *stato-acoustic nerve*

warm-blooded = *homiothermic*

SUBJECT INDEX

231

Index of Labels*